W9-BXG-821

Statistical Process Control for Quality Improvement

A Training Guide to Learning SPC

JAMES R. EVANS

*Department of Quantitative Analysis
and Information Systems*

*College of Business Administration
University of Cincinnati*

Prentice Hall, Englewood Cliffs, New Jersey 07632

Library of Congress Cataloging-in-Publication Data

Evans, James R. (James Robert), 1950-
    Statistical process control for quality improvement : a training
guide to learning SPC / James R. Evans.
      p.   cm.
    Includes bibliographical references and index.
    ISBN 0-13-558990-8
    1. Process control--Statistical methods.   I. Title.
TS156.8.E95 1991
658.5'62--dc20   91-7553
                                                            CIP

Editorial/production supervision
  and interior design: ***bookworks***
Cover design: *Lundgren Graphics, Ltd.*
Manufacturing buyers: *Kelly Behr and Susan Brunke*
Acquisitions editor: *John Willig*

 Copyright 1991 by Prentice-Hall, Inc.
A Simon & Schuster Company
Englewood Cliffs, New Jersey 07632

The publisher offers discounts on this book when ordered
in bulk quantities. For more information, write:
    Special Sales/College Marketing
    Prentice-Hall, Inc.
    College Technical and Reference Division
    Englewood Cliffs, NJ 07632

Printed in the United States of America

10 9 8 7 6 5 4 3 2 1

ISBN   0-13-558990-8

Prentice-Hall International (UK) Limited, *London*
Prentice-Hall of Australia Pty. Limited, *Sydney*
Prentice-Hall Canada Inc., *Toronto*
Prentice-Hall Hispanoamericana, S. A., *Mexico*
Prentice-Hall of India Private Limited, *New Delhi*
Prentice-Hall of Japan, Inc., *Tokyo*
Simon & Schuster Asia Pte. Ltd., *Singapore*
Editora Prentice-Hall do Brasil, Ltda., *Rio de Janeiro*

Contents

PART II: CONTROL CHARTS

PART III: PROCESS CAPABILITY ANALYSIS

PART IV: ADDITIONAL TOOLS FOR SPC

Contents

APPENDIXES

Preface

Quality is perhaps the most important issue facing manufacturers in the United States today. According to Dr. A. V. Feigenbaum, an international leader in quality management, it can be the most powerful means for achieving both customer satisfaction and lower costs. Japan and other nations have improved the quality of their products to a point where they dominate many world markets. Understanding and using methods of statistical process control (SPC) has contributed to their success.

This book is intended to be used as an introductory "nuts and bolts" training guide for developing the basic understanding and statistical skills necessary to use SPC effectively. The focus is on basic statistical tools and concepts, control charts, and process capability analysis. It is *not* intended to provide a complete overview of the quality disciplines and related managerial issues; many other good books are available for this purpose. A bibliography can be found at the end of this book. As such, this book is suited for

- production workers, technicians, and supervisors who are responsible for using SPC

- engineers without much formal statistical training who find themselves suddenly in charge of a quality function or involved in quality-related work
- managers responsible for implementing SPC who want either a refresher or a better understanding of the details.

Each chapter is focused on one major topic, and is subdivided into smaller sections. It is difficult to learn a topic such as SPC without actually doing the calculations. Within each chapter you are asked to perform various calculations, work examples, or answer questions. All answers and explanations of the answers are provided after you have had a chance to do them. To learn this material successfully, it is important that each problem and exercise *be worked!*

There is a short review quiz at the end of each section to help you see how well you have learned the material. Also, there is a comprehensive review quiz at the end of each chapter. All the answers are provided with detailed explanations in an appendix at the back of the book.

If you find that you are having trouble answering the questions, you should review the section again before moving on to the next one. You will master the fundamentals of statistical process control only with serious work. However, your efforts will be rewarded with improved quality of your operations and a positive contribution to the success of your company.

In teaching this material in industrial seminars, I have found that a slow and simple pace is essential. Even the most elementary topics require thorough step-by-step development. This book can be used for industrial training, self-study, and for vocational/community college courses.

Much of the material in this book is adapted from *MasterSPC: A Computer-Based Instructional System for Statistical Process Control,* written by the author and copyrighted by the University of Cincinnati. We thank the University of Cincinnati for permission to use this material.

James R. Evans

1

Basic Concepts of Statistical Process Control

1-1. QUALITY IMPROVEMENT—THE KEY TO THE FUTURE

In 1987, *Business Week* clearly and simply summarized the importance of quality to American industry:

> Quality. Remember it? American manufacturing has slumped a long way from the glory days of the 1950s and '60s when "Made in the U.S.A." proudly stood for the best that industry could turn out. . . . While the Japanese were developing remarkably higher standards for a whole host of products, from consumer electronics to cars and machine tools, many U.S. managers were smugly dozing at the switch. Now, aside from aerospace and agriculture, there are few markets left where the U.S. carries its own weight in international trade. For American industry, the message is simple: Get better or get beat. [June 8, 1987, p. 131]

This message is being told over and over again, even today. High-quality goods and services can give an organization a competitive edge in the international marketplace. Good quality reduces costs due to rework, scrap, returns, and warranty claims. Good quality increases productivity and profits. Most importantly, good quality generates satisfied customers who not only become repeat buyers, but spread the news of their satisfaction to many others. As a UAW vice president stated in working with Chrysler to improve quality: "No quality, no sales. No sales, no profit. No profit, no jobs."

What has caused the quality of many domestic products to fall behind that of other nations? To understand the present, and to cope with the future, we must understand the past. During the Middle Ages in Europe, the skilled craftsperson served as a designer, manufacturer, and inspector. The craftsperson knew what the customer wanted, fabricated and assembled the complete product, and took considerable pride in workmanship. There was no need for "inspection." After the industrial revolution, mass production and interchangeable parts necessitated close conformance to manufacturing specifications and made inspection an important activity. This led to the formation of separate quality departments.

Many statistical techniques for sampling inspection and process control were developed in the 1920s and 1930s at the Western Electric Company and the Bell Telephone System. Many of these tools were taught to engineers and used extensively during World War II. During the prosperity of the 1950s, when the United States was the world leader in manufacturing, these tools lost favor with manufacturers. Quality improvement was not an important issue since we had a captive market. During that time, two U.S. consultants, Drs. W. Edwards Deming and Joseph Juran, were invited to Japan to help in their rebuilding efforts. They preached the importance of quality and the use of statistical tools to improve quality. Today, we are learning from Japan and other Asian countries that we can no longer afford *not* to pay attention to quality. We have recognized that inspection alone will not result in good quality. Quality assurance is now the responsibility of *everyone* in the organization.

What do we mean by the term *quality*? The definition that has been accepted by quality professionals is "the totality of features and characteristics of a product or service that bears on its ability to satisfy given needs." This definition is the one used by the American Society for Quality Control, the nation's leading professional organization. "To satisfy given needs" means that we must be able to identify the features and characteristics of products that consumers *want*. This is often referred to as *fitness for use*. Quality must be *designed* into the product.

Once the product is designed to meet the needs of the customer, the role of manufacturing or service delivery is to ensure that the design standards and specifications are met. This is called *conformance to specifications*. Thus, we see that quality involves two major functions: quality of design and quality of conformance. Quality is *defined* by product specifications and *achieved* by production.

It is the quality of conformance that we shall be concerned with in this book. A lot can happen during manufacturing operations. Machine settings can fall out of adjustment; operators and assemblers can make mistakes; materials can be defective. Even under the most closely controlled process, there will always be variations in product output. The responsibility of production is to ensure that product specifications are met and that the final product performs as intended.

In order to know what is happening during production, we need to collect and analyze data from the process. The science of collecting, organizing, and analyzing data to draw conclusions is called *statistics*. We see and hear about statistics every day. Sports statistics such as percent pass completions, the average daily high and low temperatures, the consumer price index, and the number of automobile accidents over a holiday weekend are all examples of statistics. The only difference is that *you* will be collecting data about the process that you operate or control.

Statistics is not to be feared; in quality assurance, it does not involve complicated mathematics, only very simple calculations. While we will be working extensively with numbers, the only mathematics that you will need is basic arithmetic: addition, subtraction, multiplication, division, squares, and square roots. Appendix A provides a self-study review of these skills if you feel the need for a refresher.

When we use statistical information in a special way to monitor and control a production process, we call it *statistical process control*, or *SPC*. SPC was actually developed during the 1930s and was one of the principal tools taught to the Japanese. The Japanese used it both to control and to improve quality. All experts in quality recognize that it is an extremely important tool for quality control and improvement.

REVIEW QUIZ 1-1

_____ **1.** *True or False:* Mass production led to the formation of separate quality departments.

_____ **2.** *True or False:* "Fitness for use" means that the product conforms to design specifications.

_____ **3.** *True or False:* The responsibility of production is to ensure that design specifications are met.

_____ **4.** *True or False:* To use SPC, you will need to become an expert in statistics.

1-2. INTRODUCTION TO STATISTICAL PROCESS CONTROL

Variation, or fluctuation in performance, is all around us. In sports, for instance, no two actions are exactly alike. A professional golfer may hit a thousand identical balls with the same club, but no two will land in exactly the same spot. In fact, it is impossible to predict exactly how far a *particular* golf shot will travel. Variation in distance is due to many factors: grip, stance, muscle pressure, club speed, and so forth. Nevertheless, it is possible to predict, with a high degree of accuracy, what *percentage* of shots will go, say, between 220 and 230 yards, *if enough data on past performance are gathered and analyzed*. This is because the swing of a professional golfer is so finely tuned that every swing is nearly the same, as long as all the factors (grip, stance, and so on) remain the same. In other words, the professional golfer is *in control* of his or her swing.

Another example of variation is the life span of human beings. No one can predict exactly how long a person will live, yet insurance companies can state the pattern of life expectancy for large groups of people with very high accuracy. This is done by analyzing past data of life spans. While variation exists among individuals, the group shows stable and predictable performance.

The same concepts hold for production and manufacturing. No two outputs from any production process are exactly alike. If you measure any quality characteristic—such as the diameters of parts from a screw machine, the amount of soft drink filled in a bottle, or the number of missorted letters per day at a post office—you will find variation. Such variation is the result of many small differences in those factors that comprise a production process: people, machines, materials, methods, and measurement. For example, different lots of material will vary in composition, thickness, or moisture content; similar cutting tools will have slight variations between them; electrical fluctuations will cause variations in power which, in turn, affect the performance of equipment; and different measurement instruments will show slight differences.

It is impossible to fully understand, predict, or control all these minor variations in a production environment. They are usually called *common causes of variation*. Taken together, however, we can describe them in aggregate, just as we can describe the distances of golf shots and life spans of people as groups, though not individually. Common causes cannot be eliminated completely, but they can be reduced through changes in the process. For example, a golfer may change his or her grip or stance and thus improve consistency. Likewise, a manufacturer might purchase a new machine that produces more consistent output.

There are other causes of variation in production that can be identified and eliminated rather easily. For instance, when a tool wears down it can be replaced; when a machine falls out of adjustment, it can be reset; or when a bad lot of material is discovered, it can be returned to the supplier. These are often called *assignable causes of variation,* or *special causes of variation.* To know when any of these conditions occur, we need to collect and analyze data obtained from the process output.

A traditional approach to manufacturing is to inspect final product and screen out items that do not meet specifications. This practice is wasteful and not economical since a large amount of product may have to be scrapped or reworked if it was not manufactured correctly. A better strategy is to "do it right the first time" and not allow parts to be produced that do not meet specifications. At the very least we should be able to identify assignable causes when they occur, and adjust the process that is beginning to make parts that are out of specification before too many of them are produced. In this way, consistent quality can be assured.

We can learn a lot about a process by studying the process output. If numerical information about the process output is gathered and interpreted properly, we can better understand the ability of the process to meet specifications, and to determine when action is necessary to correct the process and prevent nonconforming product from being produced in the future.

Statistical process control (SPC) is a method of gathering and analyzing data to solve practical quality problems. The term statistical means that we will be drawing conclusions from numbers (it does not mean that we need to use a lot of mathematics). Process refers to the fact that we will be concerned with a specific production process and its ability to produce output having consistent quality. Control means that we will monitor a process and adjust it when necessary so that it will perform in the manner that it is intended to perform. To sum it up, SPC is a method for helping us monitor and control a process by gathering data about the characteristics of process output, analyzing the data, and drawing conclusions from the data.

SPC has several advantages for production operators:

1. Operators can identify problems arising from unsuitable tools, materials, fixtures, and so forth.
2. Operators can tell what variation in a process is normal and eliminate the need for constant adjustments that, in fact, can produce unwanted variation.

3. Operators can make more efficient use of their time and will produce more good product.
4. Operators can use control charts to see the progress of their work and take pride in keeping a process in control.

SPC has several advantages for managers also:

1. Managers can scientifically determine the ability of a process to produce product that meets specifications.
2. Managers can make better decisions about purchasing new equipment.
3. Managers will see lower costs and increased productivity.
4. Managers can use SPC data to provide a record to show customers the quality of a firm's products.

If both production operators and managers learn SPC, communication can be improved since they can all speak a common language through statistics. More importantly, everyone becomes involved in the quality improvement process and works toward a common goal.

REVIEW QUIZ 1-2

_____ **1.** *True or False:* Common causes of variation consist of all the minor variations in a production process that cannot be identified on an individual basis.

_____ **2.** *True or False:* Common causes of variation in production can be controlled by operators of the process.

_____ **3.** *True or False:* A good approach to quality control is to inspect the final product and to screen out items that do not meet specifications.

_____ **4.** *True or False:* Control charts allow operators to see the progress of their work and to take pride in keeping a process under control.

_____ **5.** *True or False:* SPC has many advantages, however, it does not lower cost or increase productivity.

1-3. MEASUREMENT AND STATISTICS

Data form the basis for quality control decisions and actions. Data are gathered in two principal ways: measurement and counting. Data that come from measurements along a continuous scale are called *variables data*. Examples of variables data are the length of a shaft, the inside diameter of a drilled hole, the weight of a carton, and the tensile strength of a rod. The precision with which we measure data depends on the measuring instrument. Measurements may be accurate to only two decimal places or three decimal places, for example.

Data that come from counting the number of good or bad items in a group are called *attributes data*. Examples of attributes data are the number of defective pieces in a lot, the number of surface scratches on a fender, and the percentage of broken pins in an electrical component.

When data are collected, it is important to clearly record the data, the time the data were collected, the measuring instruments that were used, who collected the data, and any other important information such as lot numbers, machine numbers, and the like. Data sheets or check sheets should be used that allow all important information to be recorded. For example, a typical data sheet for recording up to five measurements at different times might look like the one in Figure 1-1. In this figure we see

Date 4-26-89		Part Number A25B4170			
Part Name clip	**Operation** 055	**Specifications** .60 ± .10 mm			
Operator SMS	**Machine** A100	**Gage** microm.	**Unit of measurement** .00		
Time	**9:00**	**10:00**	**11:00**		
1	61	54	46		
2	64	57	48		
3	62	49	45		
4	59	53	49		
5	60	51	42		

Figure 1-1.   An example of a data collection sheet

that five measurements were recorded at 9:00, 10:00, and 11:00. By having a record of such information, we can trace the source of quality problems more easily.

Two important issues in inspection and measurement are *accuracy* and *precision*. Accuracy refers to measuring the correct value. It is defined as the amount of error in a measurement in proportion to the total size of the measurement. One measurement is more accurate than another if it has a smaller *relative error*. For example, suppose that two instruments measure a dimension whose true value is 0.250. Instrument A may read 0.248 while instrument B may read 0.259. The relative error of instrument A is the difference between the actual value and the measured value, divided by the actual value:

$$\frac{.250 - .248}{.250} = 0.008 \text{ or } 0.8 \text{ percent}$$

What is the relative error of instrument B? You may perform your calculations in the space below:

The difference between the actual value and the measurement of instrument B is .259 − .250 = .009. Dividing this by the actual value, .250, gives the relative error of .009/.250 = .036 or 3.6 percent. We would conclude that instrument A is more accurate than instrument B in this case.

Precision refers to the differences among repeated measurements. For example, if we measure the dimension using instrument A three times, we might get .248, .246, and .251. Using instrument B, we might get .259,

.258, and .259. Even though instrument B is not as accurate as instrument A, it is more precise since it measures the same value repeatedly more closely.

Accuracy and precision are important issues in using SPC since the value of any data is only as good as the data itself. If the measuring instruments are not properly calibrated, or the inspection is not conducted properly, incorrect conclusions can be drawn from the data and poor decisions will be made. Any instruments used for quality measurements must be properly calibrated and maintained in good working order.

The collection of all possible items of interest in a particular study is called a *population*. Generally, the population of interest is either so large or so unknowable that it is neither possible nor desirable to collect all the data. Instead, we use a sample. A *sample* is a portion of the population that is selected to represent the whole population. We usually call a single data value in a sample an *observation*. For example, a population might consist of 100,000 parts that are received from a supplier. We might select 500 of them to check for acceptable quality. These 500 pieces constitute the sample. As another example, a population might be defined as all parts that can be produced by a machine at a certain point in time, say 9:00 A.M. If we select the first five parts that are produced at 9:00 A.M., we have a sample.

The purpose of collecting sample data is to draw conclusions about the population. This is what the science of statistics is about and on what SPC is based. For instance, let us examine the data shown in Figure 1-1. If we look at the sample taken at 9:00, we see that all five observations fall between .59 mm and .64 mm. At 10:00, the observations range from .49 to .57. At 11:00, they fall between .42 and .49. It appears that the size of the clips is getting smaller over time.

We might suspect that this trend is due to some problem in the equipment. Statistics will help us decide whether our suspicion is true and whether to take some corrective action. More importantly, statistics can tell us when our intuition is wrong, and when to leave the process alone.

Statistical process control consists of the following:

1. selecting a sample of observations from a production process
2. measuring one or more quality characteristics
3. recording the data
4. making a few calculations

5. plotting some information on a graph called a control chart
6. examining the chart and the data to see if any unnatural problems can be identified
7. determining the cause of any problems and taking corrective action.

The determination of what quality characteristics to monitor, what types of measuring instruments to use, and what forms on which to record the data is up to the production and quality managers, so we shall not be concerned with the details of steps 1, 2, and 3 in this book. Our focus is on the *mechanics* of statistical process control. Most of this book is focused on steps 4, 5, and 6. We will learn how to make the appropriate statistical calculations, to draw control charts, and to interpret the information that appears on the charts. If any problems (that is, assignable causes of variation) are identified, it is up to the operators and supervisors to determine the causes and make adjustments. Statistics alone cannot provide this information. However, in the last chapter we shall discuss some useful quality problem-solving tools that complement SPC. Once a process is brought into control, management should attack the common causes to reduce the variation. This is the subject of *process capability,* which we also discuss later in this book.

A Note on Computers and Calculators in SPC

In many companies, implementing SPC is greatly aided by computers and PC workstations. Some excellent software exists that performs automatically all the calculations that we shall study. (The magazines *Quality* [Hitchcock Publications] and *Quality Progress* [American Society for Quality Control] publish periodic software surveys highlighting quality-related software.) Many small companies, however, particularly those just beginning SPC programs, still rely on manual construction and use of control charts. In this book we will take you step by step through the calculations that are necessary to use SPC. We encourage you to use a calculator whenever possible to ease the burden of doing these calculations. If you feel that you may need to brush up on some basic arithmetic skills (with or without a calculator), Appendix A provides a short quiz on mathematics skills and a review of basic concepts. Even if you have computer software that performs these calculations automatically, working through the calculations by hand just once will greatly improve your understanding of SPC, and we urge you to do so.

REVIEW QUIZ 1-3

_____ **1.** *True or False:* Data that come from counting are called variables data.

_____ **2.** *True or False:* The collection of all possible items of interest in a particular study is called a sample.

_____ **3.** *True or False:* A single data value in a sample is called an observation.

_____ **4.** *True or False:* SPC techniques are not only useful for identifying that a problem exists, but are also well suited to identifying the cause of the problem.

_____ **5.** *True or False:* The weight of a part is an example of variables data.

_____ **6.** *True or False:* Accuracy refers to the ability of an instrument to repeatedly measure the same value.

1-4. PLOTTING POINTS ON GRAPHS

Graphs are useful tools for displaying data. The daily newspaper usually has several examples of graphs. Figure 1-2 gives an example of a graph of the Dow Jones Industrial Averages (DJIA) which typically is found in the newspaper.

The graph in Figure 1-2 tells us at a glance that over the time period shown, the DJIA fluctuated somewhere around 2660, had a high of 2690 on the 18th, and a low of 2630 on the 22nd. Over the last five days, a downward trend is evident. Graphs are easy to draw and provide excellent visual information. It is not surprising, then, that they have extensive use in SPC.

In using control charts, you will have to plot sample observations on a similar two-dimensional graph. Such graphs consist of two axes, called

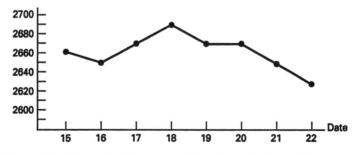

Figure 1-2.   Example of a graph of Dow Jones Industrial Averages

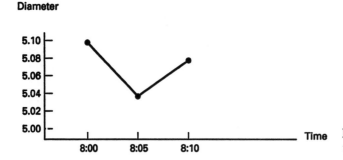

Figure 1-3. Plot of bearing diameters over time

the *horizontal axis* and the *vertical axis*. In Figure 1-2, the data axis is the horizontal axis, and the values of the DJIA are shown on the vertical axis.

On a control chart, the horizontal axis usually corresponds to the *sample number* or to the *time at which the data were collected,* and the vertical axis corresponds to a *measurement* or a *statistical calculation.* For example, suppose that we measured the diameter of one bearing selected every five minutes from a manufacturing process and found the following:

Time	Diameter
8:00	5.10
8:05	5.04
8:10	5.08

A plot of these data using time on the horizontal axis and the diameter on the vertical axis is shown in Figure 1-3. We generally connect successive points by straight lines.

REVIEW QUIZ 1-4

1. Plot the following observations collected from a production process on the chart below.

Time	Measurement
9:00	3.04
9:15	3.10
9:30	3.01
9:45	3.07
10:00	2.99
10:15	3.05

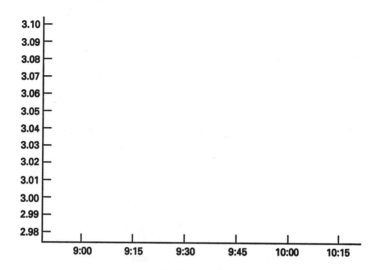

END OF CHAPTER QUIZ

_____ **1.** *True or False:* Most production processes produce output with no variation in quality characteristics.

_____ **2.** *True or False:* Inspection is a good method to control quality.

_____ **3.** *True or False:* Statistical quality control methods help us solve important quality problems by collecting and performing simple analyses of production data.

_____ **4.** An example of variables data is:
   a. the number of packages sent to the wrong address
   b. the volume of liquid filled in a bottle
   c. the percentage of broken pins in an electrical component

_____ **5.** An example of attributes data is:
   a. the number of typographical errors on a printed page
   b. the diameter of a drilled hole
   c. the weight of a carton

_____ **6.** *True or False:* A group of items selected from a population is called an observation.

_____ **7.** *True or False:* On a control chart, the horizontal axis usually represents the sample number or the time at which the data were collected.

_____ **8.** *True or False:* The purpose of statistical quality control is simply to provide information to operators and managers to help in identifying quality problems.

2

Measures of Location and Variability

In this chapter, we will discuss several important numerical measures for sample data that are used extensively in quality control. These numerical measures are often called *sample statistics* since they are computed from sample data. *Measures of location* describe the "centering" of the data; that is, they are used to represent the central value of the data. *Measures of variability* describe the "spread" of data. By spread, we mean how the data varies from the central value.

2-1. SIMPLE MEASURES OF LOCATION

Three measures that are used most often to describe the central tendency of data are the *mean, median,* and *mode*. The mean of a set of data is simply the average value of the data. Let us use a simple example to illustrate how the mean is computed. Suppose that we took measurements of the number of miles per gallon (mpg) obtained by a particular model of an automobile during a carefully controlled factory test. The values are 24.3, 24.6, 23.8, 24.0, and 24.3.

The mean, or average value, is computed by adding the values of the individual measurements and dividing this result by the number of measurements.

Step 1: Add the individual measurements:

$$
\begin{array}{r}
24.3 \\
24.6 \\
23.8 \\
24.0 \\
\underline{24.3} \\
\end{array}
$$

sum =

(Enter your answer in the space above.)
You should have found that the sum is 121.0.

Step 2: Divide the sum by the number of measurements. This gives the mean.

$$ \text{mean} = 121.0/5 = 24.2 $$

We can say that, *on the average,* this sample group of automobiles achieved 24.2 mpg, even though none of the individual automobiles obtained this figure. In statistics and quality control, it is customary to use the letter x to refer to an individual observation, and $\bar{x}$ (x-bar) to refer to the mean. So we would write: $\bar{x} = 24.2$.

The median is another statistical measure of central location of a set of data. The median is the value that falls in the middle when the data are ordered from smallest to largest. The original data are 24.3, 24.6, 23.8, 24.0, and 24.3. If we order these numbers from smallest to largest, we have:

$$ 23.8 \quad\quad 24.0 \quad\quad 24.3 \quad\quad 24.3 \quad\quad 24.6 $$

The value in the middle (the third value) is 24.3; this is the median.

The mean is most often used as a measure of central location. However, there are situations in which the median is useful. This is because the mean is affected by very large or very small values in the data, but the median is not. To illustrate this, suppose that one of the automobiles had a faulty fuel injection system and obtained only 15.8 mpg instead of the lowest value of 23.8:

$$ 15.8 \quad\quad 24.0 \quad\quad 24.3 \quad\quad 24.3 \quad\quad 24.6 $$

The median is still 24.3, but the mean is

$$ \frac{15.8 + 24.0 + 24.3 + 24.3 + 24.6}{5} = 22.6 $$

We see that the mean decreases from 24.2 to 22.6 because of the change in one value of the data. In this case, the median value of 24.3 gives a better indication of the actual mileage of this type of automobile than the mean. In general, when there are extreme values, or "outliers" (very large or very small values) in a set of data, the median usually is a better measure of central location than the mean.

If there are an even number of observations, there is no single middle value. In this case, we compute the median by taking the *average of the two middle values*. For instance, suppose we have the following six observations, ordered from the smallest to the largest:

$$3 \quad 5 \quad 9 \quad 10 \quad 15 \quad 16$$

The middle two observations are 9 and 10. The median is computed as $(9 + 10)/2 = 19/2 = 9.5$.

The final measure of central location that is used occasionally is the mode. The mode is the value that occurs with the greatest frequency. For the mpg data

$$23.8 \quad 24.0 \quad 24.3 \quad 24.3 \quad 24.6$$

we see that the value 24.3 occurs twice. Therefore, 24.3 is the mode.

In general, the mode is not as good a measure of central location as the median or the mean. However, for certain types of large data sets, the mode does provide a very quick estimate of the central value of the data. We will discuss this in the next chapter.

REVIEW QUIZ 2-1

1. The average value of a set of data is called the
    a. mean
    b. median
    c. mode

2. The middle value of a set of data is called the
    a. mean
    b. median
    c. mode

3. The most frequent observation in a set of data is called the
    a. mean
    b. median
    c. mode

4. Suppose that we have the following data:

      142   144   144   146   148   150   155

  a. What is the mean?

  b. What is the median?

  c. What is the mode?

2-2. RANGE AND STANDARD DEVIATION

Measures of location such as the mean, median, and mode provide information only about the centering of the data. They do not tell us anything about the spread, or variability, of the data. Measures of variability are needed to describe variation in a set of data.

The two most useful measures of variability for sample data in quality control are the *range* and the *standard deviation*. The range is the difference between the largest and the smallest observations in a sample. Recall the automobile mileage data that we used in the previous section:

24.3   24.6   23.8   24.0   24.3

The largest observation is 24.6 and the smallest observation is 23.8. The range is computed as

$$\text{Range} = 24.6 - 23.8 = 0.8$$

The range is often denoted by the letter R.

You see that the range is easy to compute. However, one of the disadvantages of using it is that it is influenced much by extreme values in the data. Do you remember the example in which the value 23.8 was replaced by 15.8?

Original data:	24.3	24.6	23.8	24.0	24.3
Modified data:	24.3	24.6	15.8	24.0	24.3

What is the range of the modified data?

For the modified data, the largest value is 24.6 and the smallest value is 15.8. Therefore, the range is 24.6 − 15.8 = 8.8. Observe how much the range has changed by simply changing one value. Even though four of the five observations are between 24.0 and 24.3, the presence of the extreme value 15.8 gives a very large value for the range. It would be easy to misinterpret the true variability of the data if only the range is known, especially for small samples.

Another measure of variability that involves all the observations in a data set (instead of only the smallest and largest values) is called the *standard deviation*. This is an extremely important statistical measure of variability that is used extensively in quality control. However, it is more difficult to compute than the range.

The standard deviation is a measure of the "deviation," or distance, of the observations from the mean. The deviation from the mean is the difference between the observations (x) and the mean ($\bar{x}$), or $x - \bar{x}$. Recall that in the mpg example, we calculated the mean to be $\bar{x} = 24.2$. To compute the standard deviation, we first compute the deviation from the mean for each observation. To assist us, we will use the worksheet given below. In the left column we have listed the individual values of x; in the right, we will record the deviations.

Value, x	Deviation, $x - \bar{x}$
24.3	
24.6	
23.8	
24.0	
24.3	

The first observation is 24.3. If we subtract the mean, we obtain

$$x - \bar{x} = 24.3 - 24.2 = 0.1$$

We write this in the second column of the worksheet:

Value, x	Deviation, $x - \bar{x}$
24.3	0.1
24.6	
23.8	
24.0	
24.3	

Not it's your turn. Complete this worksheet, by calculating the deviation from the mean of each remaining observation.

Your results should be:

Value, x	Deviation, $x - \bar{x}$
24.3	0.1
24.6	0.4
23.8	−0.4
24.0	−0.2
24.3	0.1

Notice that if an observation is *less* than the mean, the deviation, $x - \bar{x}$, is *negative*. What happens if you add up all the deviations? (Try it.) You find that the sum of the deviations from the mean is zero. This is *always* true, and therefore does not give us any useful information, although it is a useful check of the data to that point.

We are interested in the *distance* from the mean, regardless of which side of the mean an individual observation lies. We can eliminate the negative values of the deviations by *squaring* them. Therefore, the next step in computing the standard deviation is to square each of the deviations from the mean. For the first observation, the deviation is 0.1. The square of this number is

$$(0.1)^2 = (0.1)(0.1) = 0.01$$

We will record this in an additional column of the worksheet.

Value, x	Deviation, $x - \bar{x}$	Square of Deviation
24.3	0.1	0.01
24.6	0.4	0.16
23.8	−0.4	0.16
24.0	−0.2	0.04
24.3	0.1	0.01

We then add the squares of these deviations.
Sum of squared deviations = .01 + .16 + .16 + .04 + .01 = .38.

The final step in computing the standard deviation is to divide the sum of the squared deviations by the number of observations *minus one,* and take the square root of the result. Since there are five observations, we divide by 4:

$$.38/4 = .095$$

Finally, taking the square root, we have the standard deviation:

$$\text{standard deviation} = s = \sqrt{.095} = .30822$$

The standard deviation for sample data is usually denoted by the small letter "*s.*"

You may be wondering why the sum of the squared deviations is divided by the number of observations *minus one*. Statisticians have shown that doing so provides a more accurate estimate of the true population standard deviation when a sample is used. This is one of those times when mathematics is more important than intuition!

Summary of Computing the Standard Deviation

1. List the observations in a column.

2. Subtract the mean from each observation.

3. Square the results of step 2.

4. Add all the squared deviations.

5. Divide the sum of the squared deviations by the number of observations minus one.

6. Take the square root of the result.

Let us practice on another example. We have set up the worksheet for you.

Value, x	Deviation, $x - \bar{x}$	Square of Deviation
5		
2		
6		
4		
8		

First, compute the mean in the box below.

mean =

The mean of the data is 5.0. Now complete the worksheet. Your completed worksheet should look like the one below.

Value, x	Deviation, $x - \bar{x}$	Square of Deviation
5	0	0
2	−3	9
6	1	1
4	−1	1
8	3	9
	sum	20

Dividing by the number of observations minus one, we get

$$20/4 = 5$$

Taking the square root, we have

$$s = \sqrt{5} = 2.236$$

The formula for the standard deviation is usually written as

$$s = \sqrt{\frac{\Sigma(x - \bar{x})^2}{n - 1}}$$

The Greek letter capital sigma (Σ) stands for "sum." This means that we add up, or sum, whatever follows sigma. In this case we add the squares of the deviations from the mean. The letter n represents the number of observations. We divide the sum of the squared deviations by the number of observations minus one. Finally, we take the square root of the result.

How do we interpret the standard deviation? The data in the practice problem are repeated below, ordered from smallest to largest. Let us call this data set A.

Data set A: 2   4   5   6   8

Now let us examine another data set, which we shall call data set B:

Data set B: 1   2   5   8   9

The mean of each data set is the same, 5.0. Although the means are the same, the values in data set B are more "spread out" away from the mean than those in data set A. We can see this easily by finding the range of each data set.

What is the range of data set A?

What is the range of data set B?

The range of data set A is six, and the range of data set B is eight. Since data set B has a larger range than data set A, would you expect the standard deviation of data set B to be larger or smaller than the standard deviation of data set A? If you said larger, then you are correct. Let us verify this by computing the standard deviation for data set B.

Value, x	Deviation, $x - \bar{x}$	Square of Deviation
1	−4	16
2	−3	9
5	0	9
8	3	9
9	4	16
	sum	50

$$s = \sqrt{50/4} = \sqrt{12.5} = 3.536$$

Since the data are more spread out, the distances from the mean, $x - \bar{x}$, are larger than for data set A. When we square these numbers we get even *larger* values when compared to data set A. Therefore, the sum also is larger, and so is the standard deviation.

We can say that large values of s mean that the data are more spread out than for small values of s. A large standard deviation means there is more variability in the data than when the standard deviation is small.

REVIEW QUIZ 2-2

1. The difference between the largest and smallest observations in a data set is called the _____ .
   a. range
   b. standard deviation
   c. dispersion
2. A measure of variability that takes into account the deviations of observations from the mean is called the _____ .
   a. range
   b. standard deviation
   c. dispersion
3. What does the Greek letter sigma (Σ) stand for?
   a. sign
   b. sum
   c. standard deviation
4. What letter is used to denote the standard deviation of sample data?
5. What is the range of the following data?

$$4 \quad 3 \quad 1 \quad 7 \quad 9$$

6. Compute the standard deviation for the data in problem 5 to three decimal places. You may use the worksheet below.

Value, x Deviation, $x - \bar{x}$ Square of Deviation

END OF CHAPTER QUIZ

_____ **1.** *True or False:* Measures of location describe the spread of the data.

_____ **2.** *True or False:* The mean, median, and mode are all measures of the central location of a set of data.

_____ **3.** *True or False:* When there are extreme values in a data set (that is, very large or very small values), the median is usually a better measure of central location than the mean.

_____ **4.** *True or False:* The mode is usually as good a measure of central location as the mean and median.

_____ **5.** *True or False:* The range and standard deviation are often used to measure the "spread" or variability of a data set.

_____ **6.** *True or False:* A problem with using the range as a measure of variability is that it is influenced by extreme values.

_____ **7.** *True or False:* The range uses all data available to compute a measure of variability.

_____ **8.** *True or False:* The standard deviation uses all data available to compute a measure of variability.

_____ **9.** *True or False:* The smaller the standard deviation, the greater the variability of the data set.

3

Control Charts

3-1. INTRODUCTION TO VARIABLES CONTROL CHARTS

Suppose you were operating a machining process that drilled a hole with a specification of .500 plus or minus .003 into a part. That is, the specification is .497 to .503. The quality characteristic that is important is the inside diameter of the hole. Now suppose you measure and record the diameter of every part that is produced. You might find the following sequence of measurements for eight consecutive parts:

.501   .500   .498   .500   .499   .501   .500   .499

As we discussed in Chapter 1, there will be some variation from part to part due to materials, temperature, and so on. We can see this in these measurements. Most are close to the nominal value of .500 and within specifications.

However, suppose that after a while, you find the following measurements:

.500   .502   .503   .502   .504   .505   .504   .507

What seems to be happening? The diameters are getting larger. A plausible explanation might be that the tool is wearing down and needs replacement. This correction should be made before a large quantity of nonconforming parts are produced.

It is usually not economical or practical to inspect and measure every item from a production process, especially if a large volume is produced. An alternative is to sample parts from the process at periodic intervals to determine the state of quality.

You previously learned that there are two sources of variation in any production process: variation due to common causes that *cannot* be identified or controlled, and variation due to special causes that *can* be identified and eliminated. Tool wear is an example of a special cause. When only common causes are present, the process is said to be *in control*. When special causes are present, the process is said to be *out of control* and needs correction.

Control charts are a graphical means of identifying when special causes are present and corrective action is needed. More importantly, control charts tell us when the process is in control and *should be left alone!*

You might be asking, why not inspect a part every once in a while and if it varies from a target value, make a small adjustment to the process? The answer to this question is emphatically NO! for two reasons. First, as we have pointed out several times, some variation (due to common causes) can be expected in every process. Second, overadjustment, or "knob twiddling" of a controlled process will actually cause *more* variation than if the process were left alone. To see this, consider the following simplified example.

Suppose that the natural variation in a process is such that it produces parts whose measurements fall on a target value, say 5.0, 60 percent of the time, and are off in either direction by one-tenth the remaining 40 percent of the time. In other words, out of ten parts, six will have a value of 5.0, two will have a value of 4.9, and two will have a value of 5.1, *on the average*. Under these assumptions, if the process is left alone, *all* of the output will be between 4.9 and 5.1 as long as no assignable causes are present.

Now suppose that the operator of the process uses the following rule:

> If the value of a measurement is *below* the target value of 5.0, I will adjust the process setting *upward* by the difference; if the value of a measurement is *above* the target value of 5.0, I will adjust the target setting *downward* by the difference.

Therefore, if the operator discovers a part with a value of 4.9, for example, he or she will adjust the process setting up by 0.1, and so forth.

Let us see what would happen if the natural variation of the process would result in the following sequence of deviations away from the target setting:

target:	5.0	5.0	5.0	5.0	5.0	5.0	5.0	5.0	5.0	5.0
deviation:	−.1	+.1	0	0	−.1	0	0	0	+.1	0
actual value:	4.9	5.1	5.0	5.0	4.9	5.0	5.0	5.0	5.1	5.0

Notice that six out of ten are on target, two are one-tenth above the target, and two are one-tenth below the target, *if we leave the process alone.*

Now let us use the same sequence of deviations, but adjust the process up or down every time the value changes according to the rule given above. Assume that the initial setting is 5.0:

target:	5.0									
deviation:	−.1	+.1	0	0	−.1	0	0	0	+.1	0
actual value:	4.9									

Since the first measurement is 4.9, the operator would adjust the setting up by .1. This actually changes the process target setting to 5.1, but the operator really thinks that he or she is bringing it back to 5.0.

target:	5.0	5.1								
deviation:	−.1	+.1	0	0	−.1	0	0	0	+.1	0
actual value:	4.9	5.2								

Now, with the new (actual) process setting of 5.1, a +.1 deviation will result in an observed measurement of 5.2. The operator thinks that the adjustment is high by two-tenths, so changes the setting downward by this amount.

target:	5.0	5.1	4.9							
deviation:	−.1	+.1	0	0	−.1	0	0	0	+.1	0
actual value:	4.9	5.2	4.9							

Again, the 4.9 reading will cause the operator to adjust the setting upward by one-tenth. If he or she continues in this fashion, we find the following results:

target:	5.0	5.1	4.9	5.0	5.0	5.1	5.0	5.0	5.0	4.9
deviation:	−.1	+.1	0	0	−.1	0	0	0	+.1	0
actual value:	4.9	5.2	4.9	5.0	4.9	5.1	5.0	5.0	5.1	4.9

What is the result of all this unnecessary adjustment? Only three parts are produced that have the target value of 5.0 (30 percent); four parts are produced with a value of 4.9 (40 percent); two with a value of 5.1 (20 percent); and one with a value of 5.2 (10 percent). We see very clearly that the variation is actually greater than if the process were left alone. The only time a process should be adjusted is when special causes are identified; that is what control charts are for.

Control charts were first used by Dr. Walter Shewhart at Bell Laboratories in the 1920s (they are sometimes called *Shewhart charts*). Dr. Shewhart was the first to make a distinction between common causes of variation and special causes of variation. The control chart is the principal tool of statistical quality control.

In the remainder of this chapter we will learn how to construct, use, and interpret control charts. There are many different types of control charts. We will introduce control charts for one of the most common applications in statistical quality control: inspection and measurement of variables data. Control charts for attributes data will be studied later.

The basic procedure for constructing and using a control chart is first to inspect samples of output from a production process at regular time intervals, measure the quality characteristic(s) of interest, and record the data. It is best to use *at least* 25–30 samples to construct a control chart; however, in order not to make the following illustration too complicated, we will use a smaller sample size.

Suppose we are interested in controlling the diameter of a metal shaft. Every hour, a sample of five shafts is chosen, and the diameter is measured. The results are shown below:

Sample

1	2	3	4	5	6	7	8	9	10	11	12
.965	.975	.975	.955	.975	.960	.960	.975	.970	.960	.985	.980
.970	.985	.980	.970	.975	.970	.980	.980	.980	.970	.965	.975
.965	.970	.970	.970	.970	.975	.970	.965	.980	.965	.995	.980
.965	.985	.970	.980	.975	.985	.975	.980	.980	.980	.950	.975
.985	.965	.975	.965	.975	.970	.975	.970	.980	.965	.985	.970

Assume that the specification on the diameter is .970 ± .015, or .955 to .985. In order to maintain these specifications, we must make sure that the process remains *centered* around the nominal value of .970, and also that the *variation* in the process remains stable.

To have quantitative information on the centering and variation of the process, we compute and monitor two statistics: the *mean* of each sample, and the *range* of each sample. We will use a *subscript* to denote the sample number. For instance, x_1 is the mean of sample number 1; x_2 is the mean of sample number 2; R_8 is the range of sample number 8, and so on.

The mean of the first sample is

$$\bar{x}_1 = (.965 + .970 + .965 + .965 + .985)/5$$

$$= 4.850/5$$

$$= .970$$

The range of the first sample is

$$R_1 = .985 - .965 = .020$$

We can record these results below the observations. Most data collection forms used with control charts are set up this way.

Sample

	1	2	3	4	5	6	7	8	9	10	11	12
	.965	.975	.975	.955	.975	.960	.960	.975	.970	.960	.985	.980
	.970	.985	.980	.970	.975	.970	.980	.980	.980	.970	.965	.975
	.965	.970	.970	.970	.970	.975	.970	.965	.980	.965	.995	.980
	.965	.985	.970	.980	.975	.985	.975	.980	.980	.980	.950	.975
	.985	.965	.975	.965	.975	.970	.975	.970	.980	.965	.985	.970
sum	4.850											
avg.	.970											
range	.020											

What is the mean of sample number 2?

The mean of sample number 2 is .976. This is computed as follows:

$$\bar{x}_2 = (.975 + .985 + .970 + .970 + .975)/5 = .976$$

What is the range of sample number 2?

The range is the difference between the largest value (.985) and the smallest value (.965) in the sample, or

$$R_2 = .985 - .965 = .020$$

If we calculate the mean and range for every sample, we have the following results:

Sample

	1	2	3	4	5	6	7	8	9	10	11	12
	.965	.975	.975	.955	.975	.960	.960	.975	.970	.960	.985	.980
	.970	.985	.980	.970	.975	.970	.980	.980	.980	.970	.965	.975
	.965	.970	.970	.970	.970	.975	.970	.965	.980	.965	.995	.980
	.965	.985	.970	.980	.975	.985	.975	.980	.980	.980	.950	.975
	.985	.965	.975	.965	.975	.970	.975	.970	.980	.965	.985	.970
avg	.970	.976	.974	.968	.974	.972	.972	.974	.978	.968	.976	.976
range	.020	.020	.010	.025	.005	.025	.020	.015	.010	.020	.045	.010

The next step is to compute the *overall mean*, denoted as $\bar{\bar{x}}$ (x double bar) and the *average range*, $\bar{R}$ (R bar). The overall mean is the average of all sample means, and the average range is the average of all the ranges. To compute the overall mean, we first add all the $\bar{x}$ values in the table:

$$.970 + .976 + .974 + .968 + .974 + .972 + .972 + .974 + .978$$
$$+ .968 + .976 + .976 = 11.678$$

Then we divide this sum by the number of samples (not the sample size). The number of samples in this example is 12. Therefore, the overall mean is

$$\bar{\bar{x}} = 11.678/12 = .973 \quad \text{(to three decimal places)}$$

The average range is computed in a similar fashion. We first add all the sample ranges:

$$.020 + .020 + .010 + .025 + .005 + .025 + .020 + .015 + .010$$
$$+ .020 + .045 + .010 = .225$$

Then we divide this sum by the number of samples:

$$\overline{R} = .225/12 = .0188$$

Let us summarize what we have learned about control charts so far. To construct a control chart, we first collect at least 25 samples of output from a production process, measure the quality characteristic that is of interest, and record the data. Next, we compute the mean and range of each sample. Finally, we compute the overall mean and average range. In the next section, we shall see how to represent this information graphically and see how to determine if the process is in or out of control.

REVIEW QUIZ 3-1

1. When only common causes of variation are present in a production process, the process is said to be _____ .
   a. in control
   b. out of control
2. When the special causes of variation are present in a production process, the process is said to be _____ .
   a. in control
   b. out of control
_____ 3. *True or False:* Control charts are used to determine if common causes of variation are present.
_____ 4. *True or False:* Control charts are often called Shewhart charts.
_____ 5. *True or False:* In constructing a control chart, we take one large sample of data at the end of a production run.
_____ 6. *True or False:* In a control chart, we are interested in controlling both the centering and the variation of the process.
_____ 7. *True or False:* The overall mean is the sum of the sample means.

3-2. CONSTRUCTING VARIABLES CONTROL CHARTS

We are now ready to learn to draw control charts. For variables data we will use two control charts: one for the sample means—called the $\bar{x}$-chart ("*x*-bar" chart), and one for the sample ranges—called the *R*-chart. We will draw the $\bar{x}$-chart first and use the data on the shaft diameters that we worked with in the previous section.

The skeleton of the $\bar{x}$-chart is shown below. At the right of the chart we have displayed the sample means and ranges that we computed in the previous lesson.

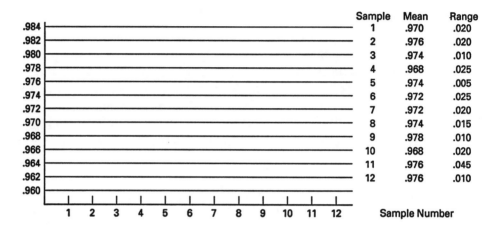

Sample	Mean	Range
1	.970	.020
2	.976	.020
3	.974	.010
4	.968	.025
5	.974	.005
6	.972	.025
7	.972	.020
8	.974	.015
9	.978	.010
10	.968	.020
11	.976	.045
12	.976	.010

The scale on the vertical axis will be used to plot the values of the sample means. The value of the overall mean should be near the center and the limits of the scale should be chosen so that the largest and smallest sample means will fall well within this range. For example, the largest $\bar{x}$ value is .978 and the smallest is .968. Both fall within the limits we have selected on the chart. A horizontal line is drawn through the overall mean, $\bar{\bar{x}}$. This is called the *center line*.

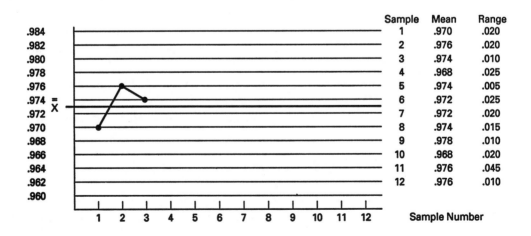

Sample	Mean	Range
1	.970	.020
2	.976	.020
3	.974	.010
4	.968	.025
5	.974	.005
6	.972	.025
7	.972	.020
8	.974	.015
9	.978	.010
10	.968	.020
11	.976	.045
12	.976	.010

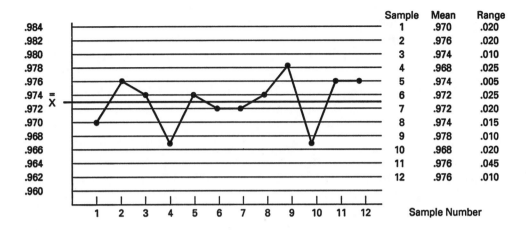

For each sample, we plot the value of $\bar{x}$ on the chart. The first three sample means are plotted below as an illustration. Take time now to plot the remaining points on the chart.

Your results should look like the chart shown above.

The $\bar{x}$-chart allows us to see how the centering of the process output varies over time. At this point we do not know whether the process is in control. We will investigate this shortly.

Next, we illustrate how to construct the R-chart in a similar manner by plotting the ranges on a new chart. On the vertical axis we plot the range of R values with $\bar{R}$ being near the center. The value of each range is then plotted on the chart. This is shown below.

There is one more step that we must perform to complete the charts. That is to compute *control limits*. Control limits are boundaries within which the process is operating in statistical control. Control limits are

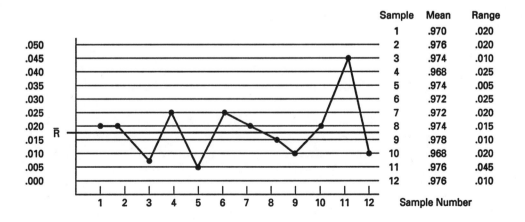

based on past performance and tell us what values we can expect for $\bar{x}$ or R *as long as the process remains stable*. If a point falls outside the control limits or if some unusual pattern occurs, then we should be suspicious of a special cause.

We always compute control limits for the R-chart first and make sure that it is in control. If it is not, then the $\bar{x}$-chart may give us misleading information. There are two control limits on each chart: the *upper control limit* (UCL) and the *lower control limit* (LCL). Control limits for the R-chart are computed by multiplying the average range, $\bar{R}$, by certain factors that are found in tables that have been developed by statisticians. The factors are different for different sample sizes, so you must be very careful to use the correct one.

A partial table of these factors is given below; Appendix C provides a more complete table.

Sample size	A_2	D_3	D_4
3	1.023	0	2.574
4	0.729	0	2.282
5	0.577	0	2.114
6	0.483	0	2.004

The upper control limit for the R-chart is given by the formula

$$UCL = D_4\bar{R}$$

This formula means that we multiply the factor D_4 corresponding to the appropriate sample size by the average range. Since the sample size is 5, $D_4 = 2.114$. Therefore, the upper control limit is

$$UCL = 2.114(.0188) = .040$$

The lower control limit for the R-chart is given by the formula

$$D_3\bar{R}$$

In this example, D_3 for a sample size of five is 0, so therefore the lower control limit is 0. We then draw and label these control limits on the chart.

If the process is in control—that is, no special causes are present—then it is very unlikely that any points will fall outside the control limits. It may happen by chance; however, there is usually only less than a 1 percent chance that this will occur.

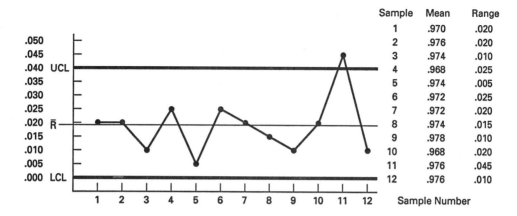

Sample	Mean	Range
1	.970	.020
2	.976	.020
3	.974	.010
4	.968	.025
5	.974	.005
6	.972	.025
7	.972	.020
8	.974	.015
9	.978	.010
10	.968	.020
11	.976	.045
12	.976	.010

If all the sample ranges are within the control limits, then we may proceed to construct the control chart for averages. If not, then you should suspect that some special causes may have been present. For example, we see that the range for sample 11 falls above the upper control limit. Upon checking production records, we might find that the regular operator was on a break and a substitute operator was running the process. This might lead us to believe that the substitute operator had made errors in measuring the shafts, or was incorrectly operating the equipment. This would be an example of a special cause, since the substitute operator is not representative of the true performance of the process. In this case, we should throw out this data and recompute the average range and the control limits. If there are still points that are out of control, then we should look for other special causes that may have been present.

On the other hand, we may find no special cause that can be explained. In this case, the point outside the control limits was simply a "freak" or a chance occurrence. We should leave it alone and proceed to construct the $\bar{x}$-chart.

If we find that several points (3 or more) fall outside the control limits, then there is some inherent instability in the process. We should investigate the process thoroughly and remove the special causes that are producing this instability. Then we should start over and collect new data from which we would construct new control charts.

In our example, suppose that sample 11 fell outside the control charts due to a special cause—the substitute operator. If we throw out this sample, we must recompute $\bar{\bar{x}}$ and $\bar{R}$. Compute new values for $\bar{\bar{x}}$ and $\bar{R}$ in the box on the next page.

You should have obtained

$$\bar{\bar{x}} = (.970 + .976 + .974 + .968 + .974 + .972 + .972 + .974 \\ + .978 + .968 + .976)/11 = .971$$

and

$$\bar{R} = (.020 + .020 + .010 + .025 + .005 + .025 + .020 + .015 \\ + .010 + .020 + .010)/11 = .0164$$

(Note that since sample 11 was eliminated, we have only 11 values in computing these averages.)

The new upper control limit is

$$UCL = 2.114(.0164) = .035$$

The revised R-chart is shown on the next page.

It is common practice to simply circle the points that are out of control and to note the reason on the chart. This also provides a useful history of the process in order to recognize specific problems that continually reoccur. The ranges now appear to be in control, so we move on to computing control limits for averages.

Control limits for the $\bar{x}$-chart are computed using the factor A_2 in the table of control limit factors. Since the sample size is 5, we see that $A_2 = .577$. The upper control limit for the $\bar{x}$-chart is given by the formula

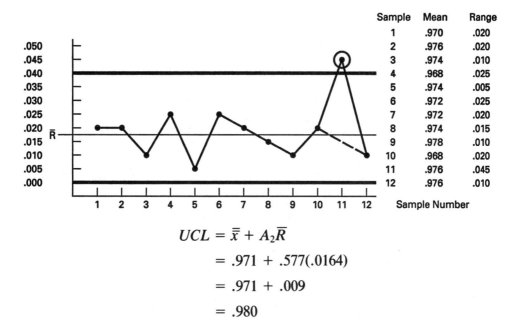

Sample	Mean	Range
1	.970	.020
2	.976	.020
3	.974	.010
4	.968	.025
5	.974	.005
6	.972	.025
7	.972	.020
8	.974	.015
9	.978	.010
10	.968	.020
11	.976	.045
12	.976	.010

$$UCL = \bar{\bar{x}} + A_2\bar{R}$$

$$= .971 + .577(.0164)$$

$$= .971 + .009$$

$$= .980$$

The lower control limit is computed by the formula

$$LCL = \bar{\bar{x}} - A_2\bar{R}$$

Find the lower control limit for this example.

You should have computed LCL to be .962. These control limits are then drawn on the chart on page 38.

The sample means appear to be in control. If we had found one or two points outside the control limits, then we would seek special causes and revise the control limits as we did for the range chart. If many points are out of control, then we would need to identify and remove special causes, collect new data, and construct new charts.

Let us summarize the process of constructing a control chart.

1. Collect at least 25 to 30 samples from a process at periodic time intervals. Samples of size 5 are commonly used in practice.

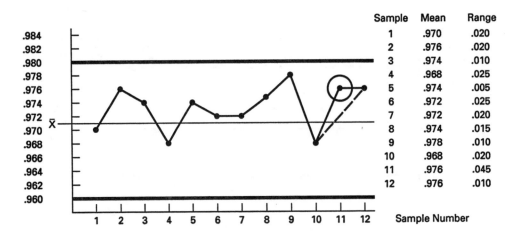

Sample	Mean	Range
1	.970	.020
2	.976	.020
3	.974	.010
4	.968	.025
5	.974	.005
6	.972	.025
7	.972	.020
8	.974	.015
9	.978	.010
10	.968	.020
11	.976	.045
12	.976	.010

2. Record the measurements on a data sheet and compute the average and range for each sample. Plot these on the charts.

3. Compute the overall mean, $\bar{\bar{x}}$, and the overall range, $\bar{R}$.

4. Using the formulas and table of control chart factors, compute control limits and draw them on the chart.

5. Check the R-chart first. If it is in statistical control, move on to the $\bar{x}$-chart. If not, determine special causes, eliminate these data points, and recompute the overall average, range, and control limits.

6. After the R-chart is brought into control, examine the $\bar{x}$-chart in a similar fashion.

Once control limits are determined, they should not be recalculated unless a fundamental change in the process is made. For example, if the purchasing department changes suppliers for the raw material used in the process, if the existing machine is replaced by a new one, or if a new employee is assigned to the process on a permanent basis, then new data should be collected and new control limits should be calculated.

Just because a process is in control does not mean that it is capable of meeting specifications on the quality characteristic that is being measured. For instance, $\bar{x}$ is a measure of the centering of the process. If the value of $\bar{\bar{x}}$ is not close to the target (nominal) mean, then the process is off-center and some adjustment would be necessary. This would not show up on the control chart alone.

Also, do not be fooled into thinking that if a process is in control, then it is producing all conforming product. Control limits *are not* specification limits and cannot be related to specification limits for sample sizes of two or more. In order to see how well a process can meet specifications, a process capability study should be conducted. We will see how to do this later in this book.

REVIEW QUIZ 3-2

_____ **1.** *True or False:* The $\bar{x}$-chart is used to plot individual observations of production output.

_____ **2.** *True or False:* In an $\bar{x}$-chart, the center line corresponds to the overall mean.

3. Which control chart should we always construct and analyze first?
   a. $\bar{x}$-chart
   b. R-chart

4. Suppose that samples of size 3 are chosen and that the average range is $\bar{R} = .84$, and the overall mean is $\bar{\bar{x}} = 2.40$.
   a. What is the upper control limit on the R-chart?

   b. What is the lower control limit on the R-chart?

   c. What is the upper control limit for the $\bar{x}$-chart?

   d. What is the lower control limit for the $\bar{x}$-chart?

_____ **5.** *True or False:* If a point falls outside a control limit, then a special cause must be present.

_____ **6.** *True or False:* Whenever several points fall outside control limits on an $\bar{x}$- or R-chart, then special causes should be investigated and new data should be collected to construct the control chart.

7. What is the correct sequence of steps in constructing control charts?
   a. Compute control limits.
   b. Compute $\bar{x}$ and R for each sample.
   c. Collect samples at periodic intervals and record the observations.
   d. Compute the overall mean and average range.
   e. Draw the control limits and center line on the chart and plot the points.

3-3. USING CONTROL CHARTS FOR CONTINUED PRODUCTION

Once a control chart has been set up, that is, the center line and upper and lower control limits have been determined, it can be used to monitor production on a continuing basis. At periodic intervals, say every hour or half-hour, we take a sample of n consecutive items from a process, and measure the quality characteristic that we wish to control. The value of n—the number of observations in each sample—must be the same as the sample size used in constructing the charts. For $\bar{x}$- and R-charts, we typically use a sample size of 5.

We will use the control charts for shaft diameters that were constructed in the previous lesson to illustrate how they can help identify and correct problems that arise during production. For the shaft diameter data, we found that the revised control limits for the $\bar{x}$-chart were

$$UCL = .980$$

$$\bar{\bar{x}} = .971$$

$$LCL = .962$$

For the R-chart, we have

$$UCL = .035$$

$$\bar{R} = .0164$$

$$LCL = 0$$

Figure 3-1 presents a typical form used for collecting data and plotting $\bar{x}$- and R-charts. These forms are available through the American Society for

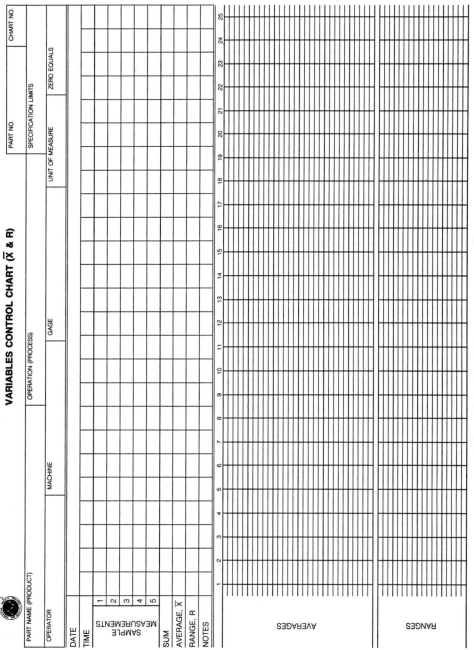

Figure 3-1. Variables control chart ($\overline{X}$ and R)

41

Quality Control. We would draw the limits on the blank chart for use in monitoring the process.

The space for recording the observations looks like the table below.

Sample	1	2	3	4	5	6	7
1							
2							
3							
4							
5							
sum							
mean							
range							

Suppose that the first sample of observations are

.972   .977   .973   .977   .971

We record these in the space provided in the first column, compute the sum, mean, and range:

Sample	1	2	3	4	5	6	7
1	.972						
2	.977						
3	.973						
4	.977						
5	.971						
sum	4.870						
mean	.974						
range	.006						

Then we plot the values of the mean and range on the appropriate charts. We see that both $\bar{x}$ and R fall within the control limits. At this point, we would continue production and collect a new sample at the next point in time.

Suppose the next six samples that are collected are shown below. You should now compute the means and ranges and plot the points on the appropriate charts.

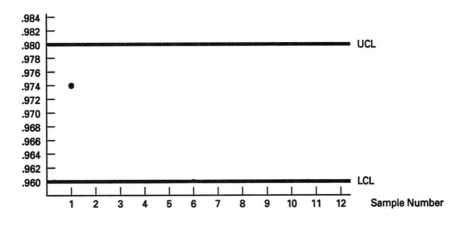

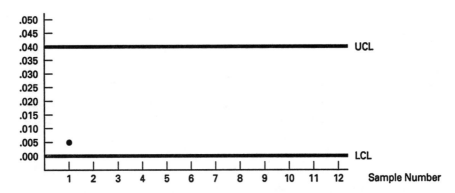

Sample	1	2	3	4	5	6	7
1	.972	.966	.971	.976	.960	.985	.986
2	.977	.960	.972	.969	.954	.970	.976
3	.973	.969	.963	.979	.963	.975	.980
4	.977	.980	.962	.977	.979	.965	.982
5	.971	.965	.962	.979	.964	.965	.986

sum	4.870	
mean	.974	
range	.006	

Your calculations should agree with the following.

Sample	1	2	3	4	5	6	7
1	.972	.966	.971	.976	.960	.985	.986
2	.977	.960	.972	.969	.954	.970	.976
3	.973	.969	.963	.979	.963	.975	.980
4	.977	.980	.962	.977	.979	.965	.982
5	.971	.965	.962	.979	.964	.965	.986
sum	4.870	4.840	4.830	4.880	4.820	4.860	4.910
mean	.974	.968	.966	.976	.964	.972	.982
range	.006	.020	.010	.010	.025	.020	.010

If you plot the means and ranges on the charts, what conclusions do you draw?

Let us examine the range chart first. In the R-chart, all the points fall between the upper and lower control limits. The points move up and down within the control limits in no definite pattern. This is expected of a process that is in control. The variation in the location of the points is due to the common causes in the process. Most importantly, these variations are beyond the control of the operator. Looking at the R-chart, we would say that the ranges are in statistical control.

Next, let us examine the $\bar{x}$-chart. Everything seems to be all right until we plot the seventh sample average. We find that it falls above the upper control limit. This tells us that there is probably a special cause affecting the process. It should be investigated and corrected now, before any further production resumes. The cause might be a broken tool, a fixture that has fallen out of adjustment, a poor quality batch of raw material, or some other reason. A control chart cannot tell you *what* the problem is, only *that* a problem exists. It is important to correct the problem before a large quantity of bad product is made. If this is not done, the control chart serves no useful purpose.

If a process is in control, it is equally important to leave it alone! Overadjustment, as we have seen earlier in this book, will only increase the variability of the process and increase the chance that poor quality products are made.

Let us now review what we have learned in this section. First, to use control charts to monitor a process, we take samples (usually five observations) at periodic intervals (every hour or so, depending on the rate of production) and measure the quality characteristic in which we are interested. Second, we record these measurements on the control chart, and compute the sample average and sample range. Third, we plot the sample averages on the x-chart and the sample ranges on the R-chart. Finally, we check to see that the points fall within the control limits. If a point falls outside a control limit, there is probably a special cause. This must be investigated and corrected before production continues. If the chart indicates that the process is in control, it should be left alone.

There is one point that we must emphasize again. *Control limits have nothing to do with specification limits!* Specifications are set by the designer; control limits are based on the variation of the process itself. Moreover, control charts describe the variation in *sample ranges* and *sample averages,* not in individual units. It is wrong to compare individual specifications with control limits based on samples.

REVIEW QUIZ 3-3

_____ **1.** *True or False:* In using a control chart to monitor production, the sample size may be different from that used to construct the chart.

_____ **2.** *True or False:* In a process that is in control, the points in a control chart should fall between the control limits with no definite pattern.

_____ **3.** *True or False:* A point falling outside a control limit probably indicates that a special cause is present in the process.

_____ **4.** *True or False:* If a special cause is identified through a control chart, the problem should be found and corrected immediately.

_____ **5.** *True or False:* Control limits for samples are the same as specification limits for individual units.

3-4. IDENTIFYING CONTROL CHART PATTERNS

When a process is in a state of statistical control, the points on a control chart should fluctuate at random between the control limits, and no recognizable patterns should exist. In the last section we saw that a point outside control limits should be checked as a possible special cause of variation. In this section, you will learn to identify other indicators of lack of control.

The following "checklist" provides a set of general rules for examining a control chart to see if a process is in control.

1. No points are outside the control limits.

2. The number of points above and below the center line is about the same.
3. The points seem to fall randomly above and below the center line.
4. There are no steady trends of points moving toward either control limit.
5. Most points, but not all, are near the center line; only a few are close to the control limits.

Figure 3-2 shows an example of a control chart for a process that is in statistical control. We will verify this using the checklist above.

---

1. Are any points outside the control limits?

Clearly there are none.

---

2. How many points are above the center line?

How many points are below the center line?

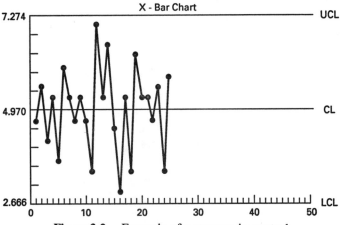

Figure 3-2.   Example of a process in control

We see that 14 of the 25 points, or 56 percent, are above the center line. This will not always be exactly 50 percent, but generally will be close to it.

When several points in a row all fall on one side of the center line, it is called a "run." An example of a run is shown in the chart in Figure 3-3. If a run occurs, this would indicate that the average value of the centering (if it is in the $\bar{x}$-chart) or the variation (if it is in the R-chart) has shifted. Typically, about 7 or 8 consecutive points, 10 out of 11, or 12 out of 14 consecutive points on one side of the center line indicates that the process has gone out of control.

---

3. Do the points seem to fall randomly above and below the center line?

This appears to be the case here. "Randomness" implies that no predictable pattern exists. This means that you cannot predict where the next point will probably fall simply by looking at the previous points. Sometimes very unusual patterns appear on a control chart. One such pattern that is not random is the cycle pattern shown in Figure 3-4. You would expect that something in the process is causing the repeated swings up and down in the value of the measured quality characteristic.

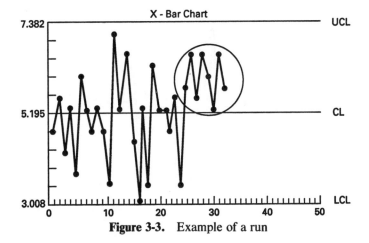

Figure 3-3.   Example of a run

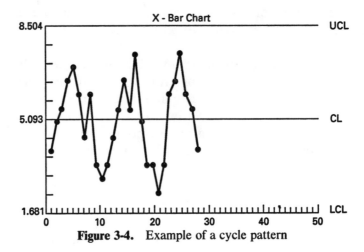

Figure 3-4.   Example of a cycle pattern

4. Is there any steady movement of points toward either control
   limit?

In Figure 3-2, there does not appear to be. A gradual movement of points
toward a control limit is called a *trend*. Trends up or down indicate that
the process is changing. An example of a trend is shown in Figure 3-5.

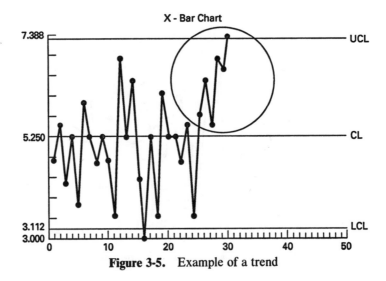

Figure 3-5.   Example of a trend

Do not be concerned with short-term trends of only a few points. This will occur quite often by chance. Of more concern are trends that show a clear pattern over a longer period of time.

---

5. Are most points, but not all, near the center line?

Are only a few points close to the control limits?

There is a very simple means of checking this. Divide the distance between the center line and each control limit into three equal parts as shown below.

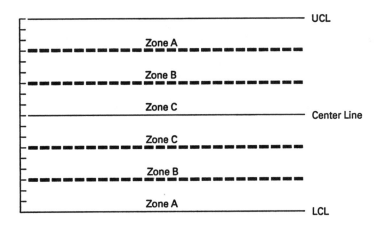

We call these areas *zones*. Zone A is the farthest from the center line, zone B is in the middle, and zone C is closest to the center line. If the process is in control, we expect to find about two-thirds of the points in zones C (combined) and only about 5 percent of them in zones A (combined).

How many points are in or on the border of zone C?

What percentage of the total points is this?

You should have counted 15 points in both of the zones C. Since there are 25 total points, this represents 60 percent. Never expect to find *exactly* two-thirds of the points in zone C, especially if you only have a small number of points in total.

How many points are in zone A?

What percentage is this?

You should have counted 2 points in zone A, representing 8 percent of the total.

You should be suspicious of a special cause if nearly all the points fall in zones C (this is called "hugging the center line") or if a large number of points fall in zone A (called "hugging the control limits").

Two simple rules are often used to help you make a decision of whether the chart is in control.

a. If 2 out of 3 consecutive points fall in zone A in one-half of the chart, then you should investigate for a special cause. This is illustrated in Figure 3-6.

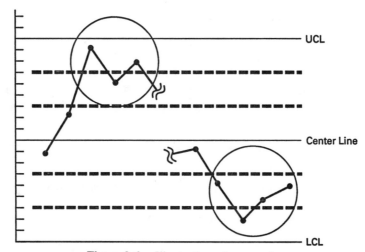

Figure 3-6.   Illustration of zone rules

b. If 4 out of 5 consecutive points fall in zones A and B in one-half of the chart, again, you should investigate. Figure 3-6 also illustrates this situation.

We will now practice using the checklist for some very simple cases. In the charts below, assume that the last 25 or so points have been in control. For the portion of the chart shown, the process is either in control or out of control. If it is out of control, one of the guidelines in the checklist is not true. You should determine which guideline is not true. To review, here is what to look for to conclude that a chart is out of control:

1. Some points are outside the control limits.
2. The number of points above the center line is either much larger or much smaller than the number of points below the center line.
3. The points do not seem to fall randomly above and below the center line.
4. You can see a steady trend of points moving toward one of the control limits.
5. Most points are near the center line, or are close to the control limits.

The answers will be given at the completion of this exercise.

PRACTICE CHART 1

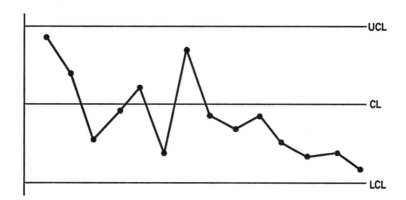

PRACTICE CHART 2

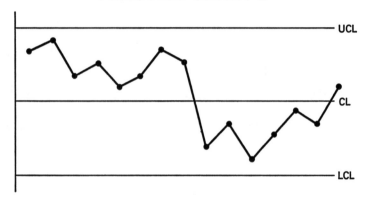

PRACTICE CHART 3

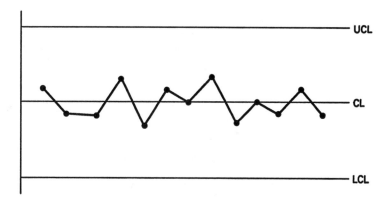

ANSWER/DISCUSSION

PRACTICE CHART 4

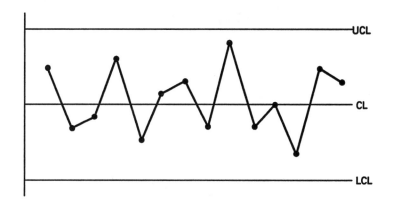

PRACTICE CHART 5

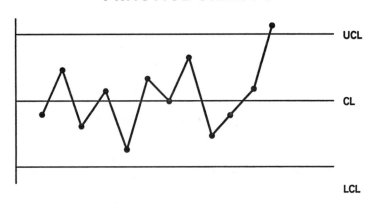

PRACTICE CHART 6

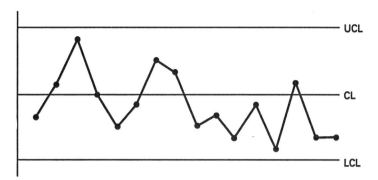

ANSWER/DISCUSSION

PRACTICE CHART 7

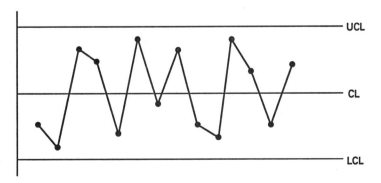

ANSWER/DISCUSSION

Discussion of Practice Charts

Practice Chart 1

Out of control. You can observe a steady trend moving toward the lower control limit. The points no longer fall randomly above and below the center line. The value of the measured quality characteristic seems to be getting gradually smaller.

Practice Chart 2

Out of control. Here, the first half of the points lie above the center line, while the second half (with the exception of the last point) lies below the center line. The points do not fall randomly above and below the center line. The value of the measured quality characteristic seems to have shifted abruptly.

Practice Chart 3

While you might conclude that the chart is in control, it appears that all the points are very close to the center line (in zone C). This is an indication that the process is out of control.

Practice Chart 4

In control.

Practice Chart 5

The last point falls above the upper control limit, the easiest out of control condition to recognize.

Practice Chart 6

Seven of the last eight points lie below the center line. It appears that the average value of the measured quality characteristic has fallen. We would conclude that this process is out of control.

Practice Chart 7

Out of control. Too many points are close to the control limits.

By now you should have a very good idea of the types of patterns that can appear on a control chart and indicate that a process has gone out of control. But what do these patterns mean? It depends on the type of chart and the specific process that is being monitored.

In general, the $\bar{x}$-chart shows where the process is centered. If the chart is in control, the process *remains* centered around the overall mean, $\bar{\bar{x}}$. If there is a trend up or down, this means that the average value of the quality characteristic is gradually moving. If there is a run on one side of the center line, then the average value has suddenly shifted. A cycle would indicate that the average value is changing in some periodic fashion. If the overall mean is the same as the nominal specification, then any of these out of control conditions would mean that the output from the process is no longer centered on the nominal specification. The centering of a process typically is affected by machine settings or some other process characteristic such as temperature, materials, or methods. Miscalibration of measurement gages can also be the problem; a shift might be indicated in the chart even though there has been no change in the actual process. Thus, it is vitally important that testing equipment be properly maintained and calibrated.

The R-chart monitors the uniformity or consistency of the process. The R-chart reacts to changes in the variation in the process. The smaller the value of R, the more uniform is the process. Any upward trend or sudden shift up in the average range is undesirable; this would mean that the variation is getting larger. Such changes are often the result of poorer materials being used, poor equipment maintenance, or fatigue of operators and inspectors, for example.

A downward shift or trend in the average range is *good!* The goal of quality assurance is to reduce the variation in a process. You should try to determine why such a change occurred and take the necessary steps to make sure that it continues.

Many unnatural patterns in the $\bar{x}$-chart are the result of a special cause in the R-chart and do not represent a true change in the centering of the process. Therefore, it is always necessary to make sure that the R-chart is in control before interpreting patterns in the $\bar{x}$-chart. Many problems in the $\bar{x}$-chart disappear once the R-chart is brought into control. We will see an example of this in the next section.

REVIEW QUIZ 3-4

_____ **1.** *True or False:* When all the points fall within the control limits, the process is said to be in control.

_____ **2.** *True or False:* If 8 of the last 10 points in a control chart are above the center line and 2 of the points are below the center line, then you should conclude that the process is out of control.

_____ **3.** *True or False:* The $\bar{x}$-chart should always be constructed and analyzed before the R-chart.

In each of the following charts, either determine that the process is in control or out of control. If you conclude that it is out of control, state the reason.

4.

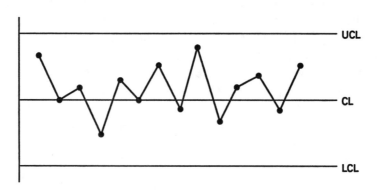

ANSWER/DISCUSSION

5.

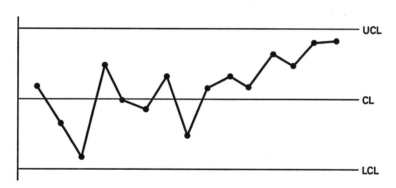

ANSWER/DISCUSSION

6.

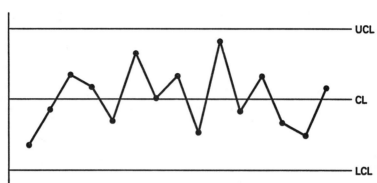

ANSWER/DISCUSSION

7.

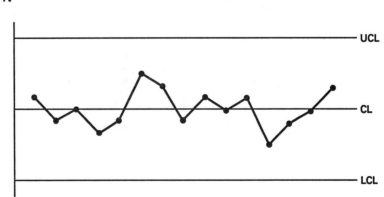

8.

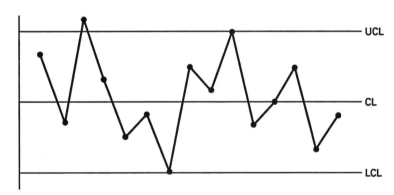

9.

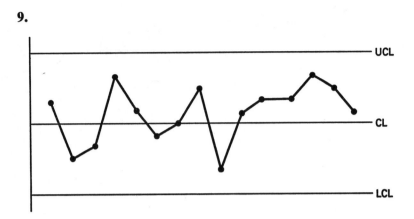

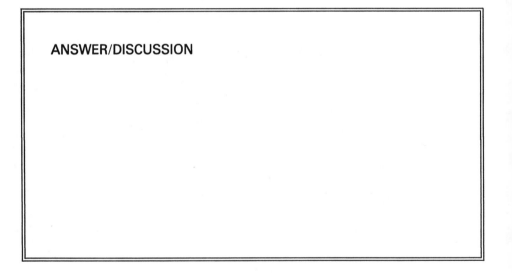

ANSWER/DISCUSSION

3-5. INTERPRETING CONTROL CHART PATTERNS

When an unusual pattern in a control chart is identified, it is up to the
operator and/or supervisor to discover the underlying special cause and
correct it. Remember that control charts can tell you only that a problem
exists; they cannot tell you what it is or how to correct it. However,
certain patterns in the $\bar{x}$- or R-charts often are the result of certain types of
causes. We caution you that these are only general guidelines designed to
help you focus on probable reasons for lack of control.

One Point Outside Control Limits

A single point outside the control limits (see Figure 3-7) is usually produced by a special cause. Often, there is a similar indication in the
R-chart. Once in a while, however, they are a normal part of the process
and occur simply by chance.

   A common reason for a point falling outside a control limit is an error
in the calculation of $\bar{x}$ or R for the sample. You should always check your
calculations whenever this occurs. Other possible causes are a sudden
power surge, a broken tool, measurement error, or an incomplete or omit-
ted operation in the process.

Sudden Shift in the Process Average

When an unusual number of consecutive points fall on one side of the
center line (see Figure 3-8), it usually indicates that the process average
has suddenly shifted. Typically, this is the result of an external influence
that has affected the process; this would be a special cause. In both the $\bar{x}$-
and R-charts, possible causes might be a new operator, a new inspector, a
new machine setting, or a change in the setup or method.

   If the shift is up in the R-chart, the process has become less uniform.
Typical causes are carelessness of operators, poor or inadequate mainte-

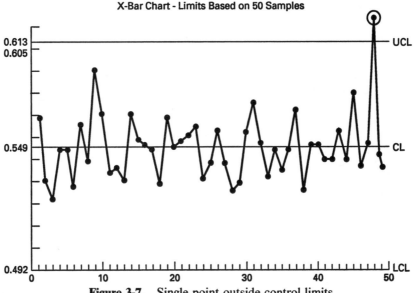

Figure 3-7.   Single point outside control limits

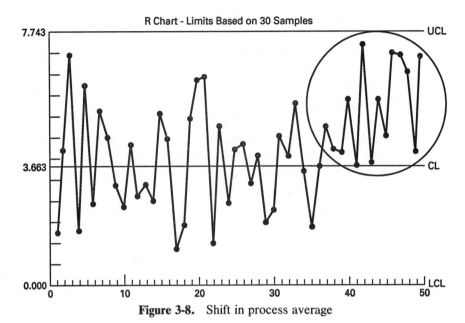

Figure 3-8.   Shift in process average

nance, or possibly a fixture in need of repair. If the shift is down in the
R-chart, uniformity of the process has improved. This might be the result
of improved workmanship, or better machines or materials. As we have
said, every effort should be made to determine the reason for the improvement and to maintain it.

Cycles

Cycles are short, repeated patterns in the chart, having alternative high
peaks and low valleys (see Figure 3-9). These are the result of causes that
come and go on a regular basis. In the $\bar{x}$-chart, cycles may be the result of
operator rotation or fatigue at the end of a shift, different gages used by
different inspectors, seasonal effects such as temperature or humidity,
differences between day and night shifts. In the R-chart, cycles can occur
from maintenance schedules, rotation of fixtures or gages, differences
between shifts, or operator fatigue.

Trends

A trend is the result of some cause that gradually affects the quality
characteristics of the product and causes the points on a control chart to
gradually move up or down from the center line (see Figure 3-10). As a

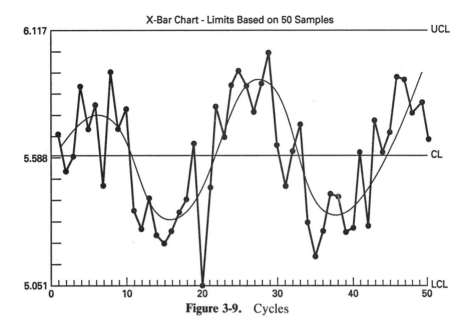

Figure 3-9.   Cycles

new group of operators gain experience on the job, for example, or as maintenance of equipment improves over time, a trend may occur. In the $\bar{x}$-chart, trends may be the result of improving operator skills, dirt or chip buildup in fixtures, tool wear, changes in temperature or humidity, or aging of equipment. In the R-chart, an increasing trend may be due to a

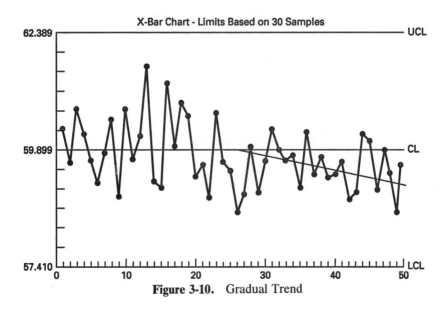

Figure 3-10.   Gradual Trend

gradual decline in material quality, operator fatigue, gradual loosening of a fixture or a tool, or dulling of a tool. A decreasing trend often is the result of improved operator skill, improved work methods, better purchased materials, or improved or more frequent maintenance.

Hugging the Center Line

Hugging the center line occurs when nearly all the points fall close to the center line (see Figure 3-11). In the control chart, it appears that the control limits are too wide. A common cause of this occurrence is when the sample is taken by selecting one item systematically from each of several machines, spindles, operators, and so on. A simple example will serve to illustrate this. Suppose that one machine produces parts whose diameters average 7.508 with variation of only a few thousandths; and a second machine produces parts whose diameters average 7.502, again with only a small variation. Taken together, you can see that the range of variation would probably be between 7.500 and 7.510, and average about 7.505. Now suppose that we sample one part from *each* machine and compute a sample average to plot on an $\bar{x}$-chart. The sample averages will consistently be around 7.505, since one will always be high and the second will always be low. Even though there is a large variation in the parts taken as whole, the sample averages will not reflect this. In such a case, it would be more appropriate to construct a control chart for *each* machine, spindle, operator, and so on.

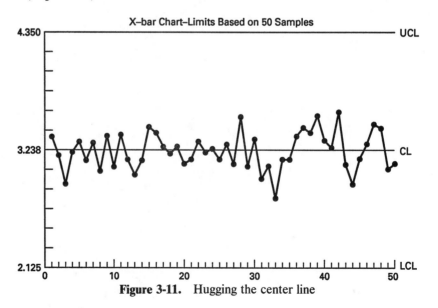

Figure 3-11. Hugging the center line

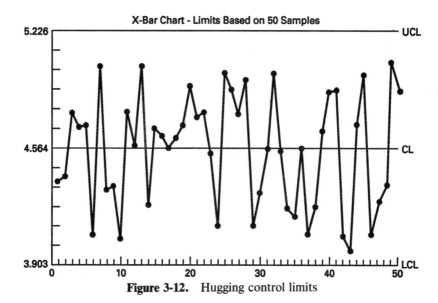

Figure 3-12.   Hugging control limits

An often overlooked cause for this pattern is miscalculation of the control limits, perhaps by using the wrong factor from the table, or misplacing the decimal point in the computations.

Hugging the Control Limits

This pattern shows up when many points are near the control limits with very few in between (see Figure 3-12). It is often called a mixture, and is actually a combination of two different patterns on the same chart. A mixture can be split into two separate patterns, as Figure 3-13 illustrates.

A mixture pattern can result when different lots of material are used in one process, or when parts are produced by different machines but fed into a common inspection group.

Instability

Instability is characterized by unnatural and erratic fluctuations on both sides of the chart over a period of time (see Figure 3-14). Points will often lie outside of both the upper and lower control limits without a consistent pattern. Assignable causes may be more difficult to identify in this case than when specific patterns are present. A frequent cause of instability is overadjustment of a machine, or the same reasons that cause hugging the control limits.

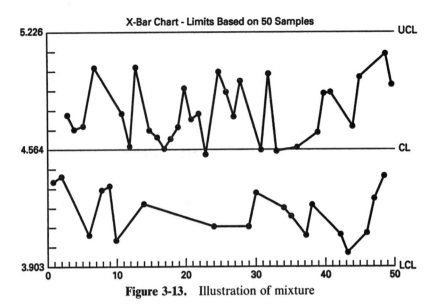

Figure 3-13. Illustration of mixture

Recall that we had stated that the *R*-chart should be analyzed before the *x̄*-chart. This is because some out of control conditions in the *R*-chart may *cause* out of control conditions in the *x̄*-chart. Figure 3-15 gives an example of this. You can see a rather drastic trend down in the range. If you examine the *x̄*-chart, you will notice that the last several points seem to be hugging the center line. As the variability in the process decreases,

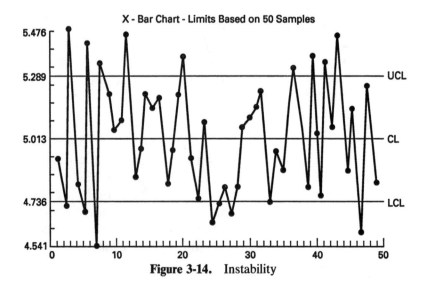

Figure 3-14. Instability

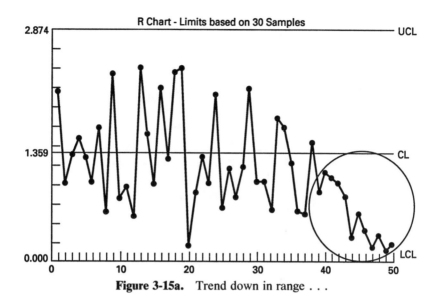

Figure 3-15a.   Trend down in range . . .

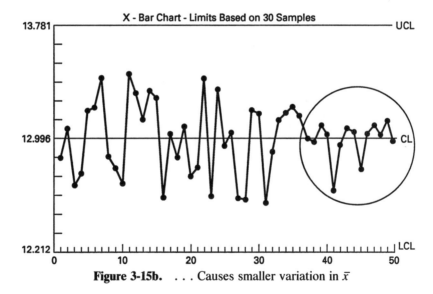

Figure 3-15b.   . . . Causes smaller variation in $\bar{x}$

all the sample observations will be closer to the true population mean, and therefore their average, $\bar{x}$, will not vary much from sample to sample. If this reduction in the variation can be identified and controlled, then new control limits should be computed for both charts.

REVIEW QUIZ 3-5

1. On the $\bar{x}$-chart, tool wear or dirt buildup will most likely result in
   a. trend
   b. single point beyond control limits
   c. mixture
   d. sudden shift of process average
   e. cycles

2. A miscalculation of $\bar{x}$ or R will often result in
   a. trend
   b. single point beyond control limits
   c. instability
   d. sudden shift of process average
   e. mixture

3. When a new machine is used for a process, it will most likely affect control charts by causing
   a. trend
   b. single point beyond control limits
   c. instability
   d. sudden shift of process average
   e. cycles

4. If a different operator is used on each of three different shifts during the day, the most likely pattern that might appear in a control chart will be
   a. trend
   b. hugging the center line
   c. instability
   d. mixture
   e. cycles

5. A miscalculation in the control limits can result in
   a. trend
   b. hugging the center line
   c. instability
   d. single point outside control limits
   e. cycles

6. Constant overadjustment of a machine may result in
   a. trend
   b. mixture
   c. instability
   d. sudden shift of process average
   e. cycles

7. If two lots of different quality material are used in the same process, the control chart pattern will probably look like

     a. trend

     b. mixture

     c. instability

     d. sudden shift of process average

     e. cycles

_____ **8.** *True or False:* A shift down in the process average in the *R*-chart means that the process has become less uniform.

_____ **9.** *True or False:* Improving operator skills or better materials will probably cause a decreasing trend in the *R*-chart.

3-6. CONTROL CHARTS FOR ATTRIBUTES

Attributes data assume only two values such as good or bad, acceptable or not acceptable, and so on. Attributes data cannot be measured, only counted. Therefore, $\bar{x}$- and *R*-charts cannot be used since they apply only to variables data. The most common control chart for attributes data is the *p-chart*. A *p*-chart monitors the proportion of nonconforming items. Sometimes it is called a *fraction nonconforming* or *fraction defective* chart.

In quality control, we distinguish between the terms *defect* and *defective*. A defect is a single nonconforming characteristic of an item. Thus, an item may have several defects. The term defective refers to an item that has one or more defects. Some charts, like the *p*-chart, are used to monitor defectives; other charts can be constructed to monitor defects.

The term *nonconforming* is more commonly used today instead of the term defective. Defective implies that something is seriously wrong with the ability of a product to function properly. A scratch may be an undesirable quality characteristic, but clearly would not affect the performance of a product. To prevent such misinterpretation, "nonconforming" is becoming more popular.

As with variables data, a *p*-chart is constructed by first gathering 25 to 30 samples of the attribute being measured. For attributes data, it is recommended that the sample size be at least 100; otherwise it is difficult to obtain good statistical results. For the type of control chart that we shall discuss, we assume that all sample sizes are the same.

Suppose that 25 samples of 100 items were inspected with a go/no-go gage. The number of nonconforming items is shown on the next page.

Sample	Number conforming
1	3
2	1
3	0
4	0
5	2
6	5
7	3
8	6
9	1
10	4
11	0
12	2
13	1
14	3
15	4
16	1
17	1
18	2
19	5
20	2
21	3
22	4
23	1
24	0
25	1

The fraction nonconforming for each sample is the number nonconforming divided by the sample size. We will call these values p_1, p_2, p_3,
and so on. Thus, p_1 is the fraction nonconforming for sample 1, p_2 is the
fraction nonconforming for sample 2, and so forth. For example, the fraction nonconforming for sample 1 is

$$p_1 = 3/100 = .03$$

What is the fraction nonconforming for sample 2?

You should have found that the fraction nonconforming for sample 2 is $1/100 = .01$. If we compute the fraction nonconforming for the remaining samples, we have the results shown below.

Sample	Number nonconforming	Fraction nonconforming
1	3	.03
2	1	.01
3	0	.00
4	0	.00
5	2	.02
6	5	.05
7	3	.03
8	6	.06
9	1	.01
10	4	.04
11	0	.00
12	2	.02
13	1	.01
14	3	.03
15	4	.04
16	1	.01
17	1	.01
18	2	.02
19	5	.05
20	2	.02
21	3	.03
22	4	.04
23	1	.01
24	0	.00
25	1	.01
		sum .55

Next, we compute the average fraction nonconforming, $\bar{p}$. This is similar to the overall mean in the $\bar{x}$-chart and the average range in the R-chart. $\bar{p}$ is the sum of the fraction nonconforming values divided by the number of samples.

$$\bar{p} = .55/25 = .022$$

This determines the location of the center line in the control chart.

Statisticians have shown that we would expect nearly all the fraction nonconforming values to fall within three standard deviations on either side of the average value, $\bar{p}$. The standard deviation is easy to compute for attributes data. The formula is

$$s = \sqrt{\frac{\bar{p}(1 - \bar{p})}{n}}$$

where n is the sample size (in this case, 100), not the number of samples. The standard deviation for the example is computed as

$$s = \sqrt{\frac{.022(1 - .022)}{100}}$$

$$= \sqrt{\frac{.022(.978)}{100}}$$

$$= \sqrt{\frac{.021516}{100}}$$

$$= \sqrt{.0002156} = .01467$$

The upper and lower control limits are given by

$$UCL = \bar{p} + 3s = .022 + 3(.01467) = .066$$
$$LCL = \bar{p} - 3s = .022 - 3(.01467) = -.022$$

Whenever LCL is negative, we use zero as the lower control limit, since the fraction nonconforming can never be negative.

We may now plot the points on a control chart just as we did for the averages and ranges. This is shown in Figure 3-16. We use the same procedures to analyze patterns in a p-chart as we did for $\bar{x}$- and R-charts. That is, we check that no points fall outside of the upper and lower control limits, and that no peculiar patterns (runs, trends, cycles, and so on) exist in the chart. The chart in Figure 3-16 appears to be in control.

Keep in mind that a p-chart shows the proportion of bad product. Changes in level in a p-chart indicate that the proportion of bad product is changing, or that we have changed our classification of "defective." A drop in level might be an indicator that an inspector fails to look for certain defects; it might not necessarily indicate an improvement in quality. One of the most important uses of a p-chart is to discover trends that indicate a deterioration or an improvement in quality.

As with variables charts, a p-chart that is in control does not necessarily mean that quality is good. An average level of defectives of 10 percent, for example, is not good quality. Management must attack the system of common causes that is responsible for this level of poor quality and make improvements in the process.

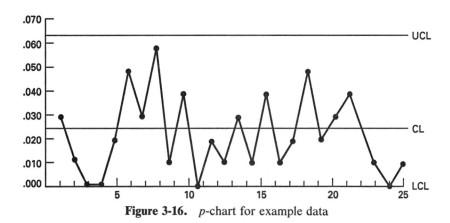

Figure 3-16. *p*-chart for example data

As you may have observed, the only differences between $\bar{x}$-, R-, and *p*-charts are the type of data that are plotted on the chart and the calculation of the control limits. The structure and the interpretation of the chart are similar in all cases. There are several other types of charts for both variables and attributes data. While the calculations differ, the basic principles are the same.

REVIEW QUIZ 3-6

_____ **1.** *True or False:* Attributes data assume only two values such as good or bad and therefore cannot be measured.

_____ **2.** *True or False:* Variables data are counted, whereas attributes data are measured.

_____ **3.** *True or False:* A control chart keeps track of the proportion of books in a sample that have missing pages. This chart would be an example of a control chart for attributes.

_____ **4.** *True or False:* A single defective item may have several defects.

5. Five samples of ten items each were selected. The number of defective items in each sample is shown below. What is $\bar{p}$?

Sample	Number of defectives
1	0
2	1
3	2
4	0
5	2

6. Suppose that $\bar{p}$ is calculated to be 0.20 and the sample size is 150. What is the standard deviation for the *p*-chart?

3-7. DESIGN ISSUES FOR CONTROL CHARTS

In designing control charts, we must consider four issues:

1. the selection of the sample data
2. the sample size
3. the frequency of sampling
4. the location of the control limits

Sampling is useful only if the sample data are representative of the entire population at the time they are taken. Each sample should reflect the system of common causes or special causes that may be present at that point in time. If special causes are present, then the sample should have a good chance of reflecting those special causes.

Consecutive measurement over a short period of time generally provide good samples for control charts, and this is the method that is most often used. (Such samples are called *rational subgroups*.) It is also important to select samples from a *single process*. A process is a specific combination of equipment, people, materials, tools, and methods. We saw that selecting samples from each of several machines can lead to hugging the control limits. Separate control charts should be set up for multiple machines or processes.

The sample size is also important. We suggested using samples of size 5; this is probably the most common in practice and the ASQC control chart forms are set up to use up to 5 observations. Small samples, however, do not allow you to detect small changes in the mean value of the quality characteristic that is being monitored. To detect small shifts in the process mean, samples of size 15 to 25 may be used. Of course, you must also consider the cost of sampling. If it is expensive or very time-consuming to take a measurement, then smaller samples may be desirable.

The third issue is the frequency of sampling. It may not be economical to sample too often. There are no hard and fast rules for sampling frequency. Samples should be close enough so that special causes can be detected before a large amount of nonconforming product is made. This decision should take into account how often special causes are observed and the volume of production.

Finally, the control limit formulas that we used in this chapter are based on three standard deviations from the mean value. This is an arbitrary decision. In Great Britain, for instance, slightly different control limits are used. In most applications, the formulas for control limits that we discussed are used. However, in certain instances you might want to consider different control limits. For example, if the cost of investigating when a point falls outside a control limit is high, then you may wish to widen the limits. If the cost of nonconforming output from the process is high, then you may wish to detect process shifts more quickly, and consequently you might use narrower control limits.

If you are unsure about any of these issues, it would be wise to consult a competent statistician who can advise you on the proper design of control charts for your applications.

END OF CHAPTER QUIZ

_____ **1.** *True or False:* If only special causes are present, the process is said to be in control.

_____ **2.** *True or False:* Control charts are a graphical means of identifying when special causes are present and corrective action is needed.

_____ **3.** *True or False:* Ten to fifteen samples is usually an adequate number of samples to take in order to construct a control chart.

4. What is the mean of the following five numbers: 5, 7, 10, 4, 4.

5. What is the range of the following five numbers: 5, 7, 10, 4, 4.

_____ **6.** *True or False:* The overall mean is the sum of the sample means.

_____ **7.** *True or False:* We should always compute control limits for the x-chart first and make sure that it is in control before analyzing the R-chart.

_____ **8.** *True or False:* In using a control chart to monitor production, the sample size may be different from the sample size used to construct the control chart.

_____ **9.** *True or False:* A point falling outside a control limit probably indicates that a special cause is present.

_____ **10.** *True or False:* There is no relationship between control limits and specification limits.

_____ **11.** *True or False:* An unusual number of points on one side of the center line, even though all points are within the control limits, is an indication that the process is out of control.

_____ **12.** *True or False:* Trends in the process average have alternate high peaks and low valleys.

_____ **13.** *True or False:* A new group of operators might be the cause of a trend.

_____ **14.** *True or False:* Hugging the center line is often caused by miscalculating the control limits.

_____ **15.** *True or False:* Overadjustment of a machine may result in instability of the process appearing on a control chart.

_____ **16.** *True or False:* $\bar{x}$- and R-charts can be used for either variables or attributes data.

_____ **17.** *True or False:* The terms defect and defective mean the same thing.

_____ **18.** *True or False:* In a p-chart, if the lower control limit is computed to be a negative number, the value of zero is used.

4

Additional Types
of Control Charts

4-1. X-BAR AND S CHARTS

X-bar and R-charts are very popular and widely used. Historically, the reason that the range was used as a measure of variability is that it is easy to compute by hand. It provides good statistical information for small sample sizes (less than 8). The standard deviation, while more difficult to compute, is a better measure of the variability of data, especially for large sample sizes. This is because it uses all of the data, whereas the range uses only two observations. An alternative to the R-chart is the *standard deviation chart,* or *s-chart.*

When the standard deviation is easy to calculate, such as when the data are recorded on a computer or if a calculator is available that has a built-in routine for computing the standard deviation, then the standard deviation chart is often used. The standard deviation chart should be used for large sample sizes.

The procedures for using $\bar{x}$- and s-charts are the same as for $\bar{x}$- and R-charts, *except* that the formulas for computing the control limits are differ-

TABLE 4-1 CONSTANTS FOR $\bar{x}$- AND $\bar{s}$-CHARTS

Sample size	B_3	B_4	A_3
2	0	3.267	2.659
3	0	2.568	1.954
4	0	2.266	1.628
5	0	2.089	1.427
6	.030	1.970	1.287
7	.118	1.882	1.182
8	.185	1.815	1.099
9	.239	1.761	1.032
10	.284	1.716	0.975

ent. For the s-chart, the upper and lower control limits are calculated as follows:

$$UCL = B_4\bar{s}$$

$$LCL = B_3\bar{s}$$

where $\bar{s}$ is the overall average standard deviation and B_3 and B_4 are constants that depend on the sample size. Table 4-1 gives a list of these constants. For the $\bar{x}$-chart, different constants must be used. The upper and lower control limits are given by:

$$UCL = \bar{\bar{x}} + A_3\bar{s}$$

$$LCL = \bar{\bar{x}} - A_3\bar{s}$$

We will illustrate the construction of an s-chart using the example from Chapter 3. The data are shown below.

Sample

	1	2	3	4	5	6	7	8	9	10	11	12
	.965	.975	.975	.955	.975	.960	.960	.975	.970	.960	.985	.980
	.970	.985	.980	.970	.975	.970	.980	.980	.980	.970	.965	.975
	.965	.970	.970	.970	.970	.975	.970	.965	.980	.965	.995	.980
	.965	.985	.970	.980	.975	.985	.975	.980	.980	.980	.950	.975
	.985	.965	.975	.965	.975	.970	.975	.970	.980	.965	.985	.970
avg	.970	.976	.974	.968	.974	.972	.972	.974	.978	.968	.976	.976
s												

We will compute the standard deviation for the first sample for you.

x	$x - \bar{x}$	$(x - \bar{x})^2$
.965	−0.005	.000025
.970	0.000	.000000
.965	−0.005	.000025
.965	−0.005	.000025
.985	0.015	.000225
		.000300

$$s = \sqrt{.0003/4} = .00866$$

What is the standard deviation for sample 2?

Your calculations should have been the following:

x	$x - \bar{x}$	$(x - \bar{x})^2$
.975	−0.001	.000001
.985	0.009	.000081
.970	−0.006	.000036
.985	0.009	.000081
.965	−0.011	.000121
		.000320

$$s = \sqrt{.00032/4} = .00894$$

If you wish to practice, you may complete the table for the remaining standard deviations. However, we suggest that you use a calculator with a built-in standard deviation function. This exercise is good practice to learn how to use that function. The results are given below.

Sample	Standard deviation
1	.00866
2	.00894
3	.00418
4	.00908
5	.00223
6	.00908
7	.00758
8	.00651
9	.00447
10	.00758
11	.01816
12	.00418

What is the overall average standard deviation?

The sum of the standard deviations is .09065. Therefore, the average standard deviation, $\bar{s}$, is .09065/12 = .00755.

To compute control limits, we see from Table 4-1 that for a sample size of 5, $B_3 = 0$, $B_4 = 2.089$, and $A_3 = 1.427$. Therefore, the control limits for the s-chart are:

$$UCL = B_4\bar{s} = 2.089(.00755) = .01577$$

$$LCL = B_3\bar{s} = 0$$

For the $\bar{x}$-chart, we have

$$UCL = \bar{\bar{x}} + A_3\bar{s} = .973 + 1.427(.00755) = .98377$$

$$LCL = \bar{\bar{x}} - A_3\bar{s} = .973 - 1.427(.00755) = .96223$$

If we draw these limits on the charts, we get the following:

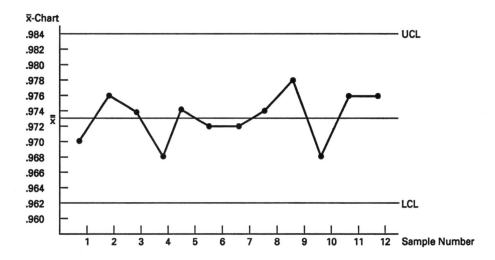

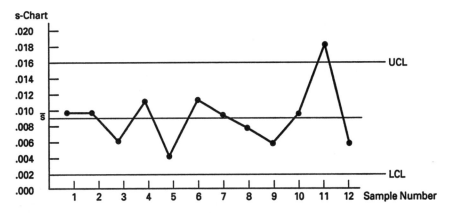

As with the R-chart, we see that sample 11 is out of control. However, it is not uncommon to find points out of control in one of the charts (R or s) that are in control in the other chart. The two charts will not always provide the same conclusions since they are based on different data.

REVIEW QUIZ 4-1

_____ **1.** *True or False:* The standard deviation chart provides better information for larger sample sizes than does the R-chart.

_____ **2.** *True or False:* The formulas for control limits in the $\bar{x}$-chart remain the same whether the R- or s-chart is used.

_____ **3.** *True or False:* The s-chart will always provide the same conclusions regarding the state of statistical control as the R-chart.

4-2. CHARTS FOR INDIVIDUALS

In using $\bar{x}$- and R- or $\bar{x}$- and s-charts, we took *samples* of observations and plotted average values on the charts. With the development of automated inspection for many processes, manufacturers can now easily inspect and measure quality characteristics on every item produced. In such cases, the sample size would be $n = 1$. Another situation in which individual measurements are more appropriate is in batch chemical processes and other low-volume production environments. If it takes several hours or days to produce one batch of chemical, for instance, waiting until a sample of 5 units is produced defeats the purpose of using a control chart. X-bar and R- or s-charts are more applicable to high-volume production operations.

In such cases, we may construct a control chart for the *individual measurements,* sometimes called x-charts (not to be confused with x-bar charts). Again, the procedures are the same; only the formulas for control limits are different. For the x-chart, the control limits are three standard deviations on either side of the overall mean $\bar{\bar{x}}$. Therefore, if you have enough historical data to obtain a good estimate of the standard deviation, the control limits would be:

$$UCL = \bar{\bar{x}} + 3s$$

$$LCL = \bar{\bar{x}} - 3s$$

An alternative is to estimate the standard deviation by using the average range. This is done as follows:

$$s = \bar{R}/d_2$$

where d_2 is a constant, shown in Table 4-2, that depends on the sample size. Using this estimate for s, the control limits would be:

$$UCL = \bar{\bar{x}} + 3\bar{R}/d_2$$

$$LCL = \bar{\bar{x}} - 3\bar{R}/d_2$$

But with samples only of size 1, how do we calculate R and monitor the variability of the process? One way is to consider *successive* observations as a larger sample and use this grouping to compute a range. This is called a *moving range*. For example, we might choose to use a moving range of 3 observations. We would group the first 3 samples and find the range of this group. Next, we would group samples 2, 3, and 4 to find the second range; then samples 3, 4 and 5, and so forth. By doing so we will have a sequence of ranges to plot on an R-chart and to compute $\bar{R}$ to estimate the standard deviation.

TABLE 4-2 CONSTANTS FOR INDIVIDUALS CHARTS

Sample size	d_2	D_3	D_4
2	1.128	0	3.267
3	1.693	0	2.574
4	2.059	0	2.282
5	2.326	0	2.114
6	2.534	0	2.004
7	2.704	.076	1.924
8	2.847	.136	1.864

We will illustrate this procedure with the following data, representing the percent of cobalt in a chemical production process.

Sample	Percent cobalt
1	3.75
2	3.80
3	3.70
4	3.20
5	3.50
6	3.05
7	3.50
8	3.25
9	3.60
10	3.10

11	4.00
12	4.00
13	3.50
14	3.00
15	3.80

What is the mean of these observations?

The mean of these data is 3.517.

Let us compute a moving range of size 2. We take the first 2 observations, 3.75 and 3.80. The range is .05.

What is the second moving range?

The second set of 2 observations is 3.80 and 3.70. Therefore, the second moving range is 0.10. If we compute the moving range for each successive pair of observations, we obtain the following.

Sample	Percent cobalt	Moving range
1	3.75	
2	3.80	.05
3	3.70	.10
4	3.20	.50
5	3.50	.30
6	3.05	.45
7	3.50	.45
8	3.25	.25
9	3.60	.35
10	3.10	.50
11	4.00	.90
12	4.00	.00
13	3.50	.50
14	3.00	.50
15	3.80	.80

What is the average range?

You should have found the average range to be 0.377. Notice that even though we have 15 observations, we only have 14 values of the range. The average range is the sum of the individual ranges divided by 14.

The upper and lower control limits for the moving range chart are given by

$$UCL = D_4\bar{R}$$

$$LCL = D_3\bar{R}$$

where D_3 and D_4 can be found in Table 4-2. From Table 4-2, for a sample size of 2, we have $D_3 = 0$, $D_4 = 3.27$, and $d_2 = 1.128$. The control limits for the moving range chart are

$$UCL = 3.27(3.77) = 1.23$$

$$LCL = 0$$

The control limits for the x-chart are

$$UCL = 3.517 + 3(.377)/1.128 = 4.520$$

$$LCL = 3.517 - 3(.377)/1.128 = 2.514$$

We leave it to you as an exercise to plot the points for these charts.

You should be somewhat cautious when using charts for individuals, however. First, the charts are not as sensitive to changes in the process as are $\bar{x}$- and R-charts; the process must vary a lot before a shift in the mean is detected. Second, the usual interpretations of the charts may be misleading. Short cycles and trends may appear on these charts that would not appear on $\bar{x}$- and R-charts. Third, there can be quite a bit of variability in the overall average and estimate of the standard deviation unless the number of observations is very large (100 or more).

REVIEW QUIZ 4-2

_____ **1.** *True or False:* We can plot individual observations on x- and R-charts by using the control limit formulas with sample sizes of 1.

_____ **2.** *True or False:* Individuals charts are not as sensitive to changes in the process mean as are $\bar{x}$- and R-charts.

_____ **3.** *True or False:* We use the moving range because there is not enough information in one observation to compute a measure of variability.

_____ **4.** *True or False:* If we know the standard deviation of individuals, we can use this to compute control limits on the individual measurements chart.

4-3. CHARTS FOR DEFECTS

Recall that *p*-charts are used to monitor the fraction nonconforming or fraction defective in a large sample. We defined a *defective* to be an item that has one or more nonconforming quality characteristics. Each of these nonconforming quality characteristics is called a *defect*. In many situations, you might be interested not in the number of items that contain defects, but in the number of defects themselves. Two new charts can be used to monitor the number of defects. These are the *c-chart* and *u-chart*. The *c*-chart monitors the total number of defects per unit *for a constant sample size*. The *u*-chart is used to control the average number of defects and is applicable to situations when the sample size is *not constant*. The *u*-chart is also useful when we are dealing with some continuous unit of measure such as area instead of individual units. Some examples would be the average number of defects per square yard of cloth, or per square foot of paper. In these cases, it does not matter what size of cloth or paper is actually sampled since we compute the statistic on a per unit basis.

Let us see how to construct a *c*-chart first. We collect at least 25 samples (of the same size) and record the total number of defects in each sample. We then compute the average number of defects by dividing the total number of defects by the total number of samples. For example, suppose we collect data on the number of surface defects on an auto body component. These might include scratches, dull spots, blisters, and so on. We are interested in only the total number. A sample of 30 autos revealed a total of 81 defects. The average number of defects, which we call $\bar{c}$, is

$$\bar{c} = 81/30 = 2.7 \text{ defects/component}$$

Control limits for the *c*-chart are easy to compute. The standard deviation of the average number of defects, $\bar{c}$, is $\sqrt{\bar{c}}$. Therefore, we expect nearly all observations to fall within $3\sqrt{\bar{c}}$ from the mean if the process is in statistical control. Thus, to compute the upper and lower control limits, we simply add or subtract $3\sqrt{\bar{c}}$ from the average number of defects:

$$UCL = \bar{c} + 3\sqrt{\bar{c}}$$

$$LCL = \bar{c} - 3\sqrt{\bar{c}}$$

Therefore, if $\bar{c} = 2.7$, $\sqrt{2.7} = 1.64$, and the control limits would be

$$UCL = 2.7 + 3(1.64) = 7.62$$

$$LCL = 2.7 - 3(1.64) = -2.22$$

TABLE 4-3 NUMBER OF DEFECTS FOUND
IN CLOTH SAMPLES

Sample	Size (sq ft)	Number of defects
1	15	3
2	11	4
3	11	5
4	15	3
5	15	6
6	11	8
7	15	10
8	11	3
9	11	2
10	15	3

Since LCL is less than zero, we use zero as the lower control limit. We plot the control limits and the number of defects for each sample on the chart and analyze it as before.

The u-chart is a little different since sample sizes may not be constant. To illustrate this, consider the data in Table 4-3, which represents the number of defects found in different size samples of cloth. The *number of defects per unit* is calculated by taking the number of defects for each sample and dividing by the sample size, in this case the square footage of the sample. For example, the number of defects per square foot for the first sample, which we call u_1, is

$$u_1 = 3/15 = .200$$

What is the number of defects per square foot for sample 2?

For sample 2, we have 4 defects divided by 11 square feet, or .364 defects per square foot. This is called u_2. Take the time now to complete the following table:

NUMBER OF DEFECTS FOUND IN CLOTH SAMPLES

Sample	Size (sq ft)	Number of defects	Defects/unit, u
1	15	3	.200
2	11	4	.364
3	11	5	
4	15	3	
5	15	6	
6	11	8	
7	15	10	
8	11	3	
9	11	2	
10	15	3	

Your results should agree with the following:

Sample	Defects/unit, u
1	.200
2	.364
3	.455
4	.200
5	.400
6	.727
7	.667
8	.273
9	.182
10	.200

The average number of defects per unit is found by adding the total number of defects and dividing by the total size of all samples.

What is the total number of defects?

The total number of defects is $3 + 4 + \ldots + 3 = 47$.

What is the total size of all samples?

The total size of all samples is 130 square feet. The average number of defects per square foot is therefore $47/130 = .3615$. This number is called $\bar{u}$.

Control limits are found in a manner similar to that used with c-charts. We add and subtract $3 \sqrt{\bar{u}/n}$ to the overall average, u.

$$UCL = \bar{u} + 3 \sqrt{\bar{u}/n}$$

$$LCL = \bar{u} - 3 \sqrt{\bar{u}/n}$$

where n is the sample size. Notice that if the size of each sample varies, *so will the control limits*. A major difference from all the charts we have seen is that each sample in a u-chart has *its own control limits*. For example, for sample 1, $n = 15$ sq ft. The upper and lower control limits for sample 1 are

$$UCL = .3615 + 3 \sqrt{.3615/15} = .8272$$

$$LCL = .3615 - 3 \sqrt{.3615/15} = -.1042 \text{ or } 0$$

What are the control limits for sample 2?

Since $n = 11$ for sample 2, the control limits are:

$$UCL = .3615 + 3 \sqrt{.3615/11} = .9053$$

$$LCL = .3615 - 3 \sqrt{.3615/11} = -.1823 \text{ or } 0$$

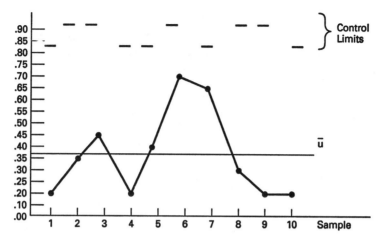

Figure 4-1. Example of u-chart

The control limits for each of the other samples will be one of these pairs since all sample sizes are either 15 or 11. Figure 4-1 shows the control chart for this example. Notice that rather than drawing one continuous straight line for the control limits, each sample has a short control limit line associated with it.

REVIEW QUIZ 4-3

_____ **1.** *True or False:* Both the c-chart and the u-chart are used for tracking the number of defectives in a process.

_____ **2.** *True or False:* In the u-chart, the control limits will vary for different samples.

_____ **3.** *True or False:* When the sample size varies, the u-chart must be used to monitor defects.

END OF CHAPTER QUIZ

1. When the sample size is large, it is better to use an
   a. *R*-chart
   b. *s*-chart

_____ **2.** *True or False:* Individuals charts are suited for low-volume production situations.

_____ **3.** *True or False:* When using an individuals chart, we can use an *R*-chart for tracking the variability.

_____ **4.** *True or False:* Defect and defective mean the same.

5. Suppose that the following sample means and standard deviations are observed for samples of size 8. Construct $\bar{x}$- and s-charts for these data.

x	s
2.15	.14
2.07	.10
2.10	.11
2.14	.12
2.18	.12
2.11	.12
2.10	.14
2.11	.10
2.06	.09
2.15	.08
2.10	.17
2.19	.13
2.14	.07
2.13	.11
2.14	.11
2.12	.14
2.08	.07
2.18	.10
2.06	.06
2.13	.14

6. Construct charts for individual measurements using both 2- and 3-period moving ranges for the following observations.

9.0   9.5   8.4   11.5   10.3   12.1   11.4   11.4   10.0   11.0   12.7   11.3   17.2

12.6   12.5   13.0   12.0   11.1   11.5   12.5   12.1

7. Construct a *c*-chart for the following data:

Sample	Number of defects
1	4
2	15
3	13
4	20
5	17
6	22
7	26
8	17
9	20
10	22

5

Frequency Distributions and Histograms

In this chapter we shall lay the foundation for process capability analysis. We need to be able to describe and summarize large groups of data. Frequency distributions and histograms are very useful tools for doing this.

5-1. SIMPLE FREQUENCY DISTRIBUTIONS

Suppose we have data on the number of automobiles sold each week for 10 weeks by an individual salesperson:

Week:	1	2	3	4	5	6	7	8	9	10
Number sold:	2	0	2	3	0	1	0	2	1	2

Looking at these data, you see that the salesperson sold either 0, 1, 2 or 3 automobiles each week. However, data that are simply listed in this way are difficult to understand and interpret. (Just imagine several hundred measurements of production output.) We need a method of summarizing the important features contained in a set of data.

A useful way of summarizing a set of data is with a *frequency distribution*. A frequency distribution is a tabular summary of a set of data showing the frequency, or number of observations, of a particular value or within a specified group. Let us examine the automobile sales data and answer a few questions about it.

How many times did the salesperson sell no automobiles?

You see that no automobiles were sold in weeks 2, 5, and 7. Therefore, the correct answer to this question is 3.

How many times was only one automobile sold?

There were two weeks (6 and 9) in which only one auto was sold.

If we order the data from low values to high values and write down the number of observations next to each value (this is called the *frequency*), we will have constructed a frequency distribution.

Number sold	Frequency
0	3
1	2
2	4
3	1
sum	10

A frequency distribution gives us information that the unorganized data do not. For example, it clearly shows that over the 10-week period, 2 automobiles were sold more often than any other number.

Number sold	Frequency
0	3
1	2
2	4 ←
3	1

The frequency distribution also shows that half the time (5 of the 10 observations), 2 or more automobiles were sold by this salesperson.

Number sold	Frequency
0	3
1	2
2	4 ←—
3	1 ←—

What *percentage* of time has this salesperson sold no automobiles?

We see that 3 weeks out of 10, no automobiles were sold. Therefore, the correct answer is 30 percent.

Frequency distributions are very useful in quality control for summarizing and displaying data in a useful fashion. An easy way to construct a frequency distribution from a large data set is to record the data in *tally form*. Let us look at another example to see how this is done.

Suppose that a hole must be drilled in a metal cylinder with a diameter specification of 0.500 ± .002, as shown in Figure 5-1. Thirty pieces were chosen and the actual diameters were measured. The results are shown below.

.499	.502	.498	.500	.500	.503
.497	.500	.501	.501	.500	.499
.502	.499	.500	.498	.501	.497
.498	.501	.501	.496	.499	.502
.500	.499	.497	.500	.498	.499

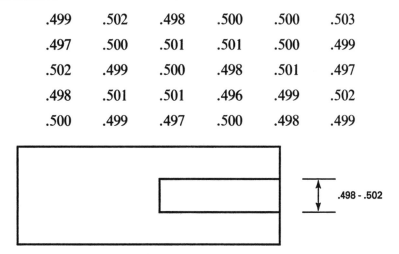

Figure 5-1. Drilled hole specifications

To construct the frequency distribution for this set of data, we first must order the data from smallest to largest. We see that the smallest measurement (observation) is .496 and the largest is .503. We list the values of all measurements in a column, from the smallest to the largest:

Measurement
.496
.497
.498
.499
.500
.501
.502
.503

We then tally each observation next to its corresponding value. The first observation is .499, so we place a tally mark (/) next to 499 in the table:

Measurement	Tally
.496	
.497	
.498	
.499	/
.500	
.501	
.502	
.503	

The second observation (down the first column) is .497, so we place a mark next to .497:

Measurement	Tally
.496	
.497	/
.498	
.499	/
.500	
.501	
.502	
.503	

We continue doing this until all observations are tallied. The final result will be:

Measurement	Tally
.496	/
.497	///
.498	////
.499	///// /
.500	///// //
.501	////
.502	///
.503	/

If we count the number of tally marks, we obtain the frequency of that measurement. The result is called the frequency distribution:

Measurement	Tally	Frequency
.496	/	1
.497	///	3
.498	////	4
.499	///// /	6
.500	///// //	7
.501	////	4
.502	///	3
.503	/	1

A frequency distribution tells us several things about the sample of 30 hole diameters. First, the data fall in the range from .496 to .503 as shown below.

Measurement	Tally	Frequency	
.496	/	1	←
.497	///	3	
.498	////	4	
.499	///// /	6	
.500	///// //	7	
.501	////	4	
.502	///	3	
.503	/	1	←

Second, 5 of the 30 diameters were not within the specifications of .498 to .502 with 4 of the 5 below the lower specification limit:

Measurement	Tally	Frequency	
.496	/	1	
.497	///	3	
.498	////	4	← lower specification limit
.499	///// /	6	
.500	///// //	7	
.501	////	4	
.502	///	3	← upper specification limit
.503	/	1	

Third, the most frequently occurring measurement was .500, with 7 of the 30 observations at this value:

Measurement	Tally	Frequency	
.496	/	1	
.497	///	3	
.498	////	4	
.499	///// /	6	
.500	///// //	7	←
.501	////	4	
.502	///	3	
.503	/	1	

Finally, about half of the measurements (14 of the 30) were .499 or below:

Measurement	Tally	Frequency	
.496	/	1	←
.497	///	3	←
.498	////	4	←
.499	///// /	6	←
.500	///// //	7	
.501	////	4	
.502	///	3	
.503	/	1	

A frequency distribution can provide useful information about what a process is doing, and enable us to compare it to what we would like it to do. This is an important first step in meeting quality requirements. For example, since 5 of the 30, or 16.67 percent, of the sample measurements are outside of the specification limits, we would expect a similar percentage to be outside the limits if we selected another sample.

If we continue to operate the process that drills holes in its current fashion, then a rather large percentage of parts will not conform to specifications. If the nonconforming parts are used, they may cause manufacturing or assembly problems later, or result in failure of the product for the customer. The only good alternative is to try to *reduce* the variation in the process to an acceptable level by improving materials, equipment, work methods, or other possible causes of variation.

Frequency distributions can also provide useful information about attributes measurements. For example, suppose you were inspecting parts for solder defects. You might use a tally sheet such as the one shown in Figure 5-2 to record your observations. Of the 41 defects observed, more than half are due to insufficient solder. Only two were due to being unsoldered. This type of analysis helps us to separate the important quality problems from the trivial ones. We would want to spend time and effort on eliminating the causes of insufficient solder first, and then move on to identifying the causes for blowholes (the next most frequent category), and so on.

This illustration is an example of Pareto's Law. Pareto's Law is generally stated as "90 percent of quality problems are caused by 10 percent of the causes." These 10 percent of the causes that result in the most quality problems are often referred to as the *vital few;* the other 90 percent of the causes are usually called the *trivial many.* We should try to identify the vital few so that corrective action can be applied where it will do the most good. Frequency distributions can easily provide this information that will lead to improved quality.

Type of defect: solder defect

Total number inspected: 50

Type	Tally	Subtotal
Blowholes	///// ///// //	12
Insufficient solder	///// ///// ///// ///// /	21
Unsoldered	//	2
Pinholes	///// /	6
	Total	41

Figure 5-2. A frequency distribution for attributes

REVIEW QUIZ 5-1

1. A tabular summary of a set of data showing the number of observations of a particular value is called a:
   a. sample
   b. frequency distribution
   c. data table

2. Suppose you have the following data:

1.325	1.326	1.324	1.328	1.326
1.326	1.325	1.328	1.327	1.325
1.325	1.326	1.326	1.327	1.327

   a. What is the smallest observation? _____
   b. What is the largest observation? _____
   c. Tally the observations in the space below.

Observation	Tally	Frequency

   d. What is the most frequent observation? _____
   e. What is the least frequent observation? _____

5-2. HISTOGRAMS

As we have seen, frequency distributions provide a useful way of summarizing a set of data. A *histogram* is a graphical representation of a frequency distribution. Recall the simple frequency distribution of the diameters of holes drilled in a metal cylinder that we introduced in Section 5-1.

Measurement	Frequency
.496	1
.497	3
.498	4
.499	6
.500	7
.501	4
.502	3
.503	1

We construct a histogram by placing the value of the individual observations on the horizontal axis of a graph and the frequencies on the vertical axis, as shown in Figure 5-3.

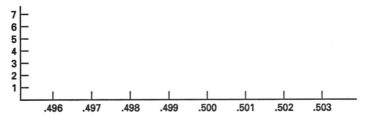

Figure 5-3.   Constructing a histogram

Next, we use a rectangle or bar whose base is centered on the value of the observation and whose height corresponds to the frequency. For example, the value .496 has a frequency of 1. We draw a bar centered on the value .496 with a height of 1 as shown below.

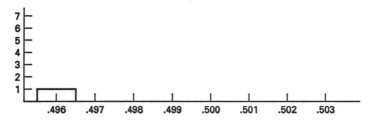

If we do this for each observation, we will have constructed the complete histogram. This is shown in Figure 5-4.

A histogram provides a convenient picture of the data. From the histogram in Figure 5-4, for example, it is easy to see that the data range from .496 to .503, and are "centered" somewhere around the values .499 or .500. We also see that the frequencies are smaller the further away we move from the center of the histogram. If the specifications on the diameters are .498 to .502, then it is very easy to see that some of the diameters are out of specification.

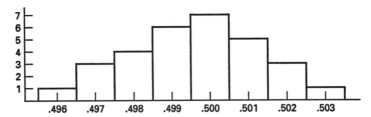

Figure 5-4. Histogram for hole diameter data

The scale on the horizontal axis that represents the values of the observations should contain no "gaps." Otherwise, the information contained in the histogram will be distorted. For example, suppose that the following frequency distribution has been constructed:

Measurement	Frequency
2.11	2
2.12	0
2.13	4
2.14	3
2.15	1

Although there are no observations with a value of 2.12, this measurement should be plotted on the histogram. See Figure 5-5. In the histogram on the left of Figure 5-5, the data do not appear to be as spread out, and erroneous conclusions might be drawn. The histogram on the right clearly shows the true distribution of the data.

Histograms may have various shapes. Some examples of different shapes of histograms are given in Figure 5-6. The shape of a histogram helps us to understand the process better. For example, in Figure 5-6(a), we see that the values of the quality characteristic that was measured are primarily clustered to the left of the histogram. Most of the measurements tend to be small, not large. In Figure 5-6(b), the opposite is found; most of the measurements tend to be large. Histograms that "tail off" to one side are called *skewed*. Thus, we would say that the histogram in Figure 5-6(a) is skewed to the right, while the histogram in Figure 5-6(b) is skewed to the left.

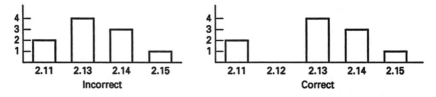

Figure 5-5. Incorrect and correct histogram representations

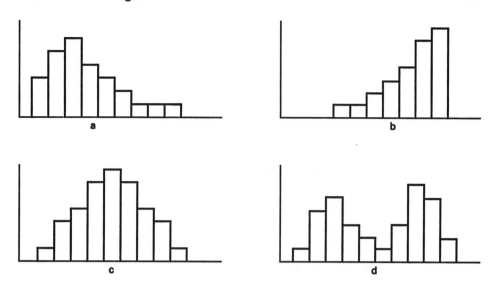

Figure 5-6. Examples of differently shaped histograms

A symmetric distribution, shown in Figure 5-6(c), means that both large and small measurements fall equally around a central value. This is typical of the output of most industrial processes. Figure 5-6(d) is called a *bi-modal* histogram because it has two distinct peaks. This indicates that there are two frequently occurring measurements. Often this results from a mixture of two different populations in the sample data. Later in this chapter we shall discuss further some applications of histograms.

REVIEW QUIZ 5-2

1. A graphical representation of a frequency distribution is called a ____ .
   a. histogram
   b. frequency chart
   c. data graph
2. The values of the observations in a simple frequency distribution or the cell limits in a grouped frequency distribution are placed on the ____ axis.
   a. horizontal
   b. vertical
3. The frequencies of the observations or cells are placed on the ____ axis.
   a. horizontal
   b. vertical
4. Which of the following is a correct histogram for the frequency distribution shown on the next page?

Value	Frequency
0	2
1	5
2	3
4	1

a.
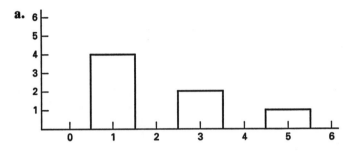

b.
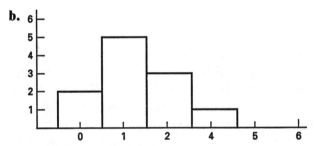

c.
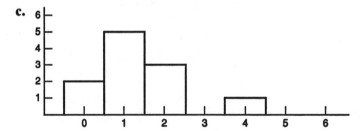

5-3 COMPUTING MEASURES OF LOCATION AND THE STANDARD DEVIATION FOR FREQUENCY DISTRIBUTIONS

While frequency distributions are useful aids for summarizing data, they do not give any precise numerical information about the data that is useful in further analyses. If we examine a frequency distribution, we find that the data typically are "centered" around some value and "spread out," or "dispersed" over some range. For example, in the frequency distribution for the hole diameters shown again below, we see that the data are centered around .499 or .500 and spread between .496 and .503. We could take the 30 individual observations and compute the mean and range or standard deviation to describe these characteristics more accurately. Fortunately, we can reduce the amount of computation that is necessary to do this.

Measurement	Tally	Frequency
.496	/	1
.497	///	3
.498	////	4
.499	///// /	6
.500	///// //	7
.501	////	4
.502	///	3
.503	/	1

When data are summarized in a frequency distribution, the mean, median, and mode can be found rather easily. To compute the mean, we need to add the values of all the observations. Remember that the frequency tells us the *number of times* that each measurement appears in the data set. So to compute the mean, we would have to add .496 once, .497 three times, .498 four times, and so forth. This is the same as *multiplying* the measurement by the frequency, and then adding the results.

Let us create a new column in the table above to help us perform this computation:

Measurement	Frequency	Measurement × Frequency
.496	1	
.497	3	
.498	4	
.499	6	
.500	7	
.501	5	
.502	3	
.503	1	
total	30	

In this new column we will multiply the measurement times its frequency, record the results, and then add them up. In the first row we write .496 times 1, or .496:

Measurement	Frequency	Measurement × Frequency
.496	1	.496
.497	3	
.498	4	
.499	6	
.500	7	
.501	5	
.502	3	
.503	1	
total	30	

In the second row, we multiply .497 by 3 to get 1.491:

Measurement	Frequency	Measurement × Frequency
.496	1	.496
.497	3	1.491
.498	4	
.499	6	
.500	7	
.501	5	
.502	3	
.503	1	
total	30	

Take a moment now to fill in the remainder of the measurement × frequency column.

Your final results should be:

Measurement	Frequency	Measurement × Frequency
.496	1	.496
.497	3	1.491
.498	4	1.992
.499	6	2.994
.500	7	3.500
.501	5	2.505
.502	3	1.506
.503	1	.503
total	30	14.987

The sum of this column would be the same as if we had added *each* of the 30 individual measurements.

To calculate the mean, we divide the sum of the (measurement times frequency) by the total number of observations. The total number of observations is equal to the sum of the frequencies. In this example, we have 30 observations. Therefore, the mean is

$$\text{mean} = \bar{x} = \frac{14.987}{30} = .500 \text{ (rounded to 3 decimal places)}$$

A general formula for the mean of a frequency distribution is

$$\text{mean} = \frac{\text{sum of all observations times their frequencies}}{\text{total number of observations}}$$

The median is easy to find in a frequency distribution since the data are already ordered from smallest to largest. The median is the middle observation. In this example, since there are 30 observations, an even number, we are looking for the fifteenth and sixteenth observations. To find these, we can add the frequencies corresponding to each value beginning with the smallest, until we find the measurement in which the fifteenth and sixteenth observations lie. These are called *cumulative frequencies*. For the first measurement, the cumulative frequency is just the frequency, 1:

Measurement	Frequency	Cumulative frequency
.496	1	1
.497	3	
.498	4	
.499	6	
.500	7	
.501	5	
.502	3	
.503	1	
total	30	

For the second row, the cumulative frequency is the frequency in the first row, 1, plus the frequency in the second row, 3:

Measurement	Frequency	Cumulative frequency
.496	1	1
.497	3	$1 + 3 = 4$
.498	4	
.499	6	
.500	7	
.501	5	
.502	3	
.503	1	
total	30	

For the next row, we add the cumulative frequency in the previous row, 4, to the frequency in that row, 4:

Measurement	Frequency	Cumulative frequency
.496	1	1
.497	3	1 + 3 = 4
.498	4	4 + 4 = 8
.499	6	
.500	7	
.501	5	
.502	3	
.503	1	
total	30	

We continue in this manner until we find the first cumulative frequency that is larger than the number of the middle observation, 16:

Measurement	Frequency	Cumulative frequency
.496	1	1
.497	3	1 + 3 = 4
.498	4	4 + 4 = 8
.499	6	8 + 6 = 14
.500	7	14 + 7 = 21 [stop here]
.501	5	
.502	3	
.503	1	
total	30	

Notice that the fifteenth and sixteenth observations are both .500. Therefore, the median is .500.

Remember that the mode is the most frequent observation.

Can you find the mode in this frequency distribution?

If you said that the mode is .500, you are correct! The value of .500 has the largest frequency (7). This, by definition, is the mode. Do not get confused in thinking that the mode is 7. The mode is the value of the *observation* that occurs with the highest frequency, *not* the value of the *frequency*.

When data are summarized in a frequency distribution, there is a short-cut method for computing the standard deviation. The formula is shown below.

$$s = \sqrt{\frac{\Sigma fx^2 - n\bar{x}^2}{n - 1}}$$

This formula says that we multiply the frequency (f) times the square of each observation x^2 and add them up. Then we multiply the number of observations (n) by the square of the mean $\bar{x}^2$. We then subtract this value from Σfx^2. Finally, we divide by the number of observations minus one, and take the square root of the result.

This is not really a very complicated procedure if we organize the calculations into a worksheet like we did for the standard deviation of individual observations. We strongly recommend that you check all the calculations on a calculator for practice.

The frequency distribution of the hole diameters is shown below.

x	f
.496	1
.497	3
.498	4
.499	6
.500	7
.501	5
.502	3
.503	1
	30 = n

First, we square each observation, x, and record the result in the new column below. For example, the square of the first observation is

$$(.496)^2 = (.496)(.496) = .246016$$

x	f	x^2
.496	1	.246016
.497	3	.247009
.498	4	.248004
.499	6	.249001
.500	7	.250000
.501	5	.251001
.502	3	.252004
.503	1	.253009
	30	

Next, we multiply the frequency f by these squared observations, record the result in the column labeled fx^2 below, and add them up.

x	f	x^2	fx^2
.496	1	.246016	.246016
.497	3	.247009	.741027
.498	4	.248004	.992016
.499	6	.249001	1.494006
.500	7	.250000	1.750000
.501	5	.251001	1.255005
.502	3	.252004	.756012
.503	1	.253009	.253009
	30		7.487091

We need the mean in order to complete the calculation of the standard deviation. You should remember that for frequency distributions, the mean is computed as the sum of the frequencies times the observations, divided by the total number of observations. We can write this as

$$\bar{x} = \frac{\Sigma fx}{n}$$

We can add an extra column for the frequencies times the observations in our worksheet.

x	f	x^2	fx^2	fx
.496	1	.246016	.246016	.496
.497	3	.247009	.741027	1.491
.498	4	.248004	.992016	1.992
.499	6	.249001	1.494006	2.994
.500	7	.250000	1.750000	3.500
.501	5	.251001	1.255005	2.505
.502	3	.252004	.756012	1.506
.503	1	.253009	.253009	.503
	30		7.487091	14.987

The mean is $14.987/30 = .4995666667$

We now have all the information we need to compute the standard deviation. First take the sum of fx^2 and subtract $n\bar{x}^2$:

$$7.487091 - 30(.4995666667)^2$$

$$= 7.487091 - 30(.2495668544)$$

$$= 7.487091 - 7.487005633$$

$$= .0000853667$$

Next, we divide by the number of observations minus one, or $30 - 1 = 29$:

$$.0000853667/29 = .0000029437$$

Finally, we take the square root:

$$s = \sqrt{.0000029437} = .0017157$$

Be assured that this will be the most complicated computation that you will ever have to make in statistical process control analysis! If you do not have access to a computer or SPC software for doing this, many calculators have built-in functions that calculate the standard deviation automatically. We recommend that you obtain and learn to use one of these.

It is important to note that serious errors can be made if you do not carry out the computations to a sufficient number of decimal places. For example, if we round all our calculations to three decimal places, we would have the following:

x	f	x^2	fx^2	fx
.496	1	.246	.246	.496
.497	3	.247	.741	1.491
.498	4	.248	.992	1.992
.499	6	.249	1.494	2.994
.500	7	.250	1.750	3.500
.501	5	.251	1.255	2.505
.502	3	.252	.756	1.506
.503	1	.253	.253	.503
	30		7.487	14.987

We would get

$$\bar{x} = .500$$

and

$$s = \sqrt{(7.487 - 30(.500)^2)/29}$$

$$= \sqrt{(7.487 - 30(.250)/29}$$

$$= \sqrt{(7.487 - 7.500)/29}$$

$$= \sqrt{-.013/29}$$

which requires us to take the square root of a negative number. This, of course, does not exist.

In Chapter 2 we briefly discussed the meaning of the standard deviation. In analyzing frequency distributions and histograms, the standard deviation has a very important application in quality control. Consider the histogram for the hole diameter data shown again below.

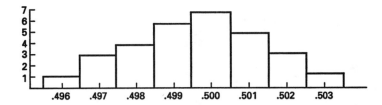

As we have stated before, the distribution of output of most production processes will follow a symmetric "bell-shaped" pattern like the one shown in the histogram. If this is so, then we can say that nearly all the observations—as much as 99.7 percent—will fall *within three standard deviations on either side of the mean.* If our sample data are representative of the whole population, then we can predict, with a high degree of confidence, the actual variation in the quality characteristic that we are measuring.

Since the mean of this distribution is approximately .500, in this example we would expect all the observations to fall between .500 − 3(.0017157) and .500 + 3(.0017157), or between .495 and .505. This is exactly what we observed in the histogram. We would believe this to be true *as long as the process is in statistical control.*

REVIEW QUIZ 5-3

1. Compute the mean and the standard deviation for the frequency distribution shown below by completing the remaining three columns in the worksheet and using the formulas that you have studied.

x	f	x^2	fx^2	fx
.5	10			
1.5	20			
2.5	60			
3.5	70			
4.5	40			

2. Suppose we have the frequency distribution shown below. Compute the mean.

value	frequency
5.0	5
6.0	3
7.0	6
8.0	1

3. What is the median of the frequency distribution in problem 1?

4. What is the mode in the frequency distribution in problem 1?

5-4. GROUPED FREQUENCY DISTRIBUTIONS

In many situations, there are too many possible values of the data to allow you to tally each individual observation. In these cases, we usually *group* the data into *classes* or *cells* that do not overlap but include all the data. If the number of observations within each class is counted, we have a *grouped frequency distribution.*

For example, suppose that we record the number of gallons of gasoline purchased by customers during a day. These values might range from only a few gallons to over 20, and include fractions of gallons. There would be far too many observations to tally. As an alternative, we may group the data into classes, for instance:

1. less than 5 gallons
2. at least 5 gallons but less than 10 gallons
3. at least 10 gallons but less than 15 gallons
4. at least 15 gallons but less than 20 gallons
5. 20 gallons or more

If we examine the data, we might find that 14 customers purchased less than 5 gallons; 86 customers bought at least 5 but less than 10 gallons; 193 customers purchased at least 10 but less than 15 gallons; 79 customers bought at least 15 but less than 20 gallons; and that 43 persons purchased between 20 and 25 gallons. We may summarize this information in the form of a grouped frequency distribution shown below:

Cell	Gallons of gasoline	Frequency
1	0 but less than 5	14
2	5 but less than 10	86
3	10 but less than 15	193
4	15 but less than 20	79
5	20 or more	43
	total	415

A grouped frequency distribution simply gives a tally of the number of observations that fall between the limits of each cell. We read from the grouped frequency distribution that in cell 2, for instance, 86 customers purchased at least 5 but less than 10 gallons of gasoline.

Each group or cell is defined by a *lower cell limit* and an *upper cell limit*. For example, in cell 1, the lower cell limit is 0 and the upper cell limit is 5. We also see that the upper cell limit of cell 1 is the lower cell limit for cell 2. But notice that cell 1 *does not include* the value 5. So if a customer pumped exactly 5 gallons of gasoline, this observation would be tallied in cell 2, not in cell 1.

We will now show how to construct a grouped frequency distribution. Suppose that the following 30 measurements were collected for some quality characteristic:

12.642	12.564	12.645	12.428	12.513	12.560
12.531	12.585	12.581	12.593	12.516	12.469
12.537	12.601	12.410	12.520	12.604	12.634
12.461	12.683	12.559	12.461	12.528	12.537
12.562	12.462	12.562	12.545	12.432	12.537

There are too many individual measurements, and each appears only once or twice. A simple frequency distribution for the individual measurements would not tell us very much about the data. However, a grouped frequency distribution would tell us a lot more about the data.

In preparing a grouped frequency distribution, we first select lower and upper limits so that all the data will fall between them. Usually we choose "nice" round numbers. Since the smallest value is 12.410 and the largest value is 12.683, we will set the lower limit as 12.400 and the upper limit as 12.700.

Next we have to determine the number of cells to have. For the moment, suppose we pick 6 cells. One rule that you should follow is to make each cell have the same width. The width of the cells can be computed easily by subtracting the lower limit from the upper limit, and dividing by the number of cells.

$$\text{cell width} = \frac{\text{upper limit} - \text{lower limit}}{\text{number of cells}}$$

Since the upper limit is 12.700 and the lower limit is 12.400, and we are using 6 cells, we determine the cell width as

$$\text{cell width} = \frac{12.700 - 12.400}{6} = \frac{0.300}{6} = 0.050$$

If we add the cell width to the lower limit, we create the upper limit of the first cell. Since the lower limit is 12.400 and the cell width is 0.050, the upper limit of the first cell is $12.400 + 0.050 = 12.450$. So the first cell will include all the observations between 12.400 and *up to but not including* the upper limit 12.450. The lower limit of the second cell will then be 12.450 and the upper limit will be $12.450 + 0.50 = 12.500$. The six cells would be as shown below:

Cell	Measurement
1	12.400 but less than 12.450
2	12.450 but less than 12.500
3	12.500 but less than 12.550
4	12.550 but less than 12.600
5	12.600 but less than 12.650
6	12.650 but less than 12.700

(If all the data are accurate to three decimal places, then we could set the upper limit of the first cell to 12.449 to avoid any confusion. We can only do this if there are no observations between 12.449 and 12.450. This would be easier if a machine operator were collecting and tallying the data. For this discussion, however, we shall simply use the cell limits as breakpoints between cells.)

Now let us go back to the actual data:

12.642	12.564	12.645	12.428	12.513	12.560
12.531	12.585	12.581	12.593	12.516	12.469
12.537	12.601	12.410	12.520	12.604	12.634
12.461	12.683	12.559	12.461	12.528	12.537
12.562	12.462	12.562	12.545	12.432	12.537

> The first observation is 12.642. In which cell does this value belong?

The observation 12.642 falls between 12.600 and 12.650, therefore it belongs in cell 5.

> The second observation is 12.531. In which cell does this value belong?

The value 12.531 belongs in cell 3, since it falls between 12.500 and 12.550.
    Now tally all the remaining observations in the workspace below and compute the frequency of each cell.

Cell	Measurement	Tally	Frequency
1	12.400 but less than 12.450		
2	12.450 but less than 12.500		
3	12.500 but less than 12.550	/	
4	12.550 but less than 12.600		
5	12.600 but less than 12.650	/	
6	12.650 but less than 12.700		

You should have obtained the grouped frequency distribution given below.

Cell	Measurement	Frequency
1	12.400 but less than 12.450	3
2	12.450 but less than 12.500	4
3	12.500 but less than 12.550	9
4	12.550 but less than 12.600	8
5	12.600 but less than 12.650	5
6	12.650 but less than 12.700	1·

From this grouped frequency distribution, we can make certain statements about the data. First, the interval with the most frequently occurring measurements is 12.500 to 12.550. There are 9 observations in this cell. Second, over half the observations (17 out of 30) lie between 12.500

and 12.600. Third, about half the observations (16 out of 30) have values of at least 12.400 but less than 12.550.

We could have selected a smaller number of cells having larger intervals. For instance, we might have used three intervals. The width of each interval would be calculated as

$$\text{cell width} = \frac{12.700 - 12.400}{3} = \frac{0.300}{3} = .100$$

This would give us the following grouped frequency distribution:

Cell	Measurements	Frequency
1	12.400 but less than 12.500	7
2	12.500 but less than 12.600	17
3	12.600 but less than 12.700	6

Such a grouping provides less information about the nature of the data than the original 6 cells. For example, it would be difficult to determine what percentage are outside specifications, if the specifications are 12.450 to 12.650.

Generally, you should use the following guidelines to determine the proper number of cells to use.

Total number of observations	Number of cells
0–9	4
10–24	5
25–49	6
50–89	7
90–189	8
more than 189	9 or 10

We see that with 30 observations, 6 cells was the proper number to use for the example.

To construct a histogram for a grouped frequency distribution, we plot the lower and upper cell limits on the horizontal axis, and draw the bars to the appropriate height between these limits. An example is given below.

Cell	Value	Frequency
1	0 up to 5	50
2	5 up to 10	100
3	10 up to 15	200
4	15 up to 20	150
5	20 up to 25	100

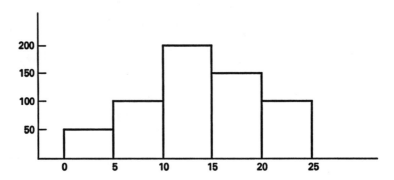

REVIEW QUIZ 5-4

1. A frequency distribution in which data are tallied into cells is called a ___ frequency distribution.
   a. simple
   b. average
   c. grouped

2. Suppose the upper limit for a grouped frequency distribution is 8.0 and the lower limit is 3.0. If 4 cells will be used, what should the cell width be?

3. Suppose we collect the following data.

2.60	2.05	2.77	2.46	2.49	2.21
2.89	2.19	2.42	2.93	2.38	2.68
2.76	2.93	2.55	2.10	2.41	2.50
2.53	2.22	2.34	2.87	2.59	2.80

   a. How many cells would be most appropriate for this data?

b. Tally the observations using the following cells:

Cell	Measurement	Tally	Frequency
1	2.00 but less than 2.20		
2	2.20 but less than 2.40		
3	2.40 but less than 2.60		
4	2.60 but less than 2.80		
5	2.80 but less than 3.00		

5-5. COMPUTING STATISTICAL MEASURES FOR GROUPED FREQUENCY DISTRIBUTIONS

As we have seen, it is often convenient to present data in a *grouped frequency distribution*. However, if the individual observations are available, we recommend computing statistical measures using the procedures discussed in the last section. When data are grouped, we lose information about the individual values. Using a grouped frequency distribution only gives *approximate* values for the mean, median, and mode. However, for large amounts of data, such approximations are usually very good, and much easier to compute if only a calculator is available.

Shown below is the grouped frequency distribution that we used in the previous section.

Cell	Measurement	Frequency
1	12.400 to 12.450	3
2	12.450 to 12.500	4
3	12.500 to 12.550	9
4	12.550 to 12.600	8
5	12.600 to 12.650	5
6	12.650 to 12.700	1
	total	30

Since each cell corresponds to a *range* of actual values, we need a single value to represent the observations in that cell. The best way to do this is to use the *midpoint* of that cell. The midpoint of any cell is found by averaging the upper and lower cell limits. For example, the midpoint of cell 1 is

$$\frac{12.400 + 12.450}{2} = \frac{24.850}{2} = 12.425$$

Let us add this information to the frequency distribution table:

Cell	Measurement	Midpoint	Frequency
1	12.400 to 12.450	12.425	3
2	12.450 to 12.500	12.475	4
3	12.500 to 12.550	12.525	9
4	12.550 to 12.600	12.575	8
5	12.600 to 12.650	12.625	5
6	12.650 to 12.700	12.675	1
		total	30

Once this is done, we can compute the mean just like an ordinary frequency distribution by using the midpoints as representative values of the cells. We first multiply the midpoints by their corresponding frequencies and then add them up:

Cell	Measurement	Midpoint	Frequency	Midpoint × Frequency
1	12.400 to 12.450	12.425	3	37.275
2	12.450 to 12.500	12.475	4	49.900
3	12.500 to 12.550	12.525	9	112.725
4	12.550 to 12.600	12.575	8	100.600
5	12.600 to 12.650	12.625	5	63.125
6	12.650 to 12.700	12.675	1	12.675
		total	30	376.300

The mean is computed by dividing the sum of the midpoints times frequencies by the total number of observations.

$$\text{mean} = \bar{x} = 376.300/30 = 12.543$$

Let us consider the median next. There are 30 observations, so the median falls between the fifteenth and sixteenth observations. We find the cumulative frequencies until we reach the sixteenth observation or higher.

Cell	Measurement	Midpoint	Frequency	Cumulative Frequency
1	12.400 to 12.450	12.425	3	3
2	12.450 to 12.500	12.475	4	7
3	12.500 to 12.550	12.525	9	16 [stop here]
4	12.550 to 12.600	12.575	8	
5	12.600 to 12.650	12.625	5	
6	12.650 to 12.700	12.675	1	

Since the cumulative frequency of cell 3 is 16, we see that the median falls between 12.500 and 12.550.

As with simple frequency distributions, the mode is determined by the cell with the largest frequency. In this example, we see that cell 3 has the largest frequency, 9. Since we use the midpoint as the representative value of the cell, the mode is 12.525.

We can use the same formula for computing the standard deviation for grouped frequency distributions. To do this, we simply use the midpoint of each cell as the "observation," x. We will illustrate this using the grouped frequency distribution example from the previous chapter.

Cell	Measurement	Midpoint	f	x^2	fx^2	fx
1	12.400 to 12.450	12.425	3	154.380625	463.1418750	37.275
2	12.450 to 12.500	12.475	4	155.625625	622.5025000	49.900
3	12.500 to 12.550	12.525	9	156.875625	1411.8806250	112.725
4	12.550 to 12.600	12.575	8	158.130625	1265.0450000	100.600
5	12.600 to 12.650	12.625	5	159.390625	796.9953125	63.125
6	12.650 to 12.700	12.675	1	160.655675	160.6556750	12.675
			30		4720.17875	376.3

We first compute the mean:

$$\bar{x} = \Sigma fx/n = 376.300/30 = 12.5433333$$

Next, we apply the formula for the standard deviation. We suggest that you verify these calculations yourself.

$$s = \sqrt{\frac{\Sigma fx^2 - n\bar{x}^2}{n - 1}}$$

$$= \sqrt{\frac{4720.17857 - 30(12.5433333)^2}{29}}$$

$$= \sqrt{\frac{4720.1875 - 4720.055333}{29}}$$

$$= \sqrt{\frac{.1311667}{29}}$$

$$= \sqrt{.0045229886} = .0672531681$$

REVIEW QUIZ 5-5

1. Compute the mean for the grouped frequency distribution shown below.

Cell	Measurement	Frequency
1	2.00 to 2.20	2
2	2.20 to 2.40	5
3	2.40 to 2.60	8
4	2.60 to 2.80	4
5	2.80 to 3.00	5

2. For problem 1, what cell contains the median? How would you estimate the median?

3. What is the mode for the data in problem 1?

4. Compute the standard deviation for the data in problem 1.

5-6 APPLICATIONS OF HISTOGRAMS

In quality control applications, a histogram is often called a *lot plot*. If the specifications of the quality measurement are drawn on the lot plot, it can tell quite a bit about the capability of the process to meet specifications. We saw this in the example earlier in this chapter. If, for instance, the specifications are .498 to .502, we can show this on the histogram by drawing vertical lines on either side of these specifications (see Figure 5-7). It is easy to see that the process cannot meet the specifications all the time.

On the other hand, suppose the specifications are .495 to .504 (see Figure 5-8). In this case we see that all the observations fall within the specification limits. We would conclude that the process is able to produce parts of acceptable quality, provided, of course, that it remains in control.

Let us examine some other lot plots and see what they can tell us about the quality of a process. Figure 5-9 provides an example that illustrates a nonconforming product being produced below the lower specification limit, but well within the upper specification limit. The process would be able to hold the specification if the nominal, or centering of the process, was set correctly. These data indicate that the centering of the process should be adjusted to the right. This will allow all of the process output to fall within the specification limits.

In Figure 5-10 we see that there are no observations below the lower specification limit. We would expect the distribution of quality measurements to gradually "tail off" on both sides of the distribution. What do

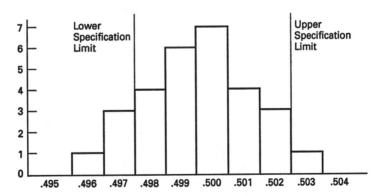

Figure 5-7. Histogram of drilled hole diameters with specifications .498–.502

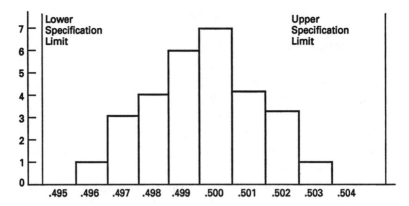

Figure 5-8. Histogram of drilled hole diameters with specifications .495–.504

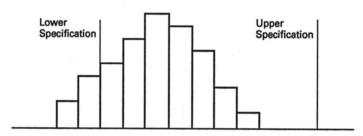

Figure 5-9. An example lot plot

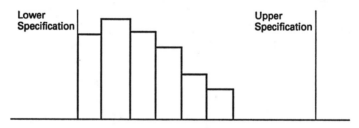

Figure 5-10. An example lot plot

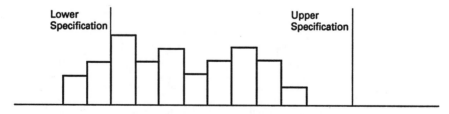

Figure 5-11. An example lot plot

you think happened? This type of lot plot probably indicates that the centering of the process is set correctly, but that all nonconforming product was sorted out of the lot.

Another situation that is often seen is shown in Figure 5-11. It appears that two different lots have been mixed together, and one of them does not meet specifications. This might occur when materials from two different suppliers are mixed, or when output from two different machines or processes are mixed together.

To see how lot plots can help in making important quality decisions, suppose that a company was planning to purchase a part that was particularly difficult to machine. The purchasing department ordered a small lot from each supplier and all the parts were inspected. The lot plot of the critical dimension of parts from supplier 1 is shown in Figure 5-12(a). It appears that supplier 1 can produce good quality parts that meet the buyer's specifications.

The lot plot for supplier 2 is shown in Figure 5-12(b). What would you conclude? While all the purchased parts meet specifications, it looks as though supplier 2 had sorted out nonconforming product before shipment. This tells the buyer that supplier 2 is probably not able to produce the part within the specifications, and that some nonconforming parts will likely be shipped. If supplier 2 were given the order, then additional inspection would be necessary to ensure that good product is always received. It appears that supplier one's process can meet the specifications

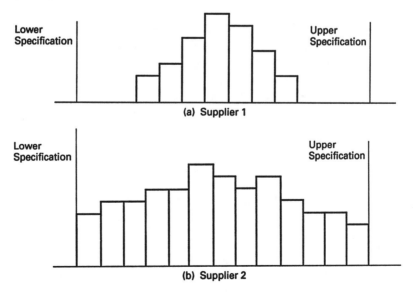

Figure 5-12.   An example lot plot

without sorting out nonconforming parts; therefore, supplier 1 will have the competitive advantage.

REVIEW QUIZ 5-6

_____ 1. *True or False:* A lot plot can show if a process needs adjustment to meet specifications.

_____ 2. *True or False:* If a lot plot does not gradually tail off on both ends, it probably means that nonconforming product was sorted out first.

_____ 3. *True or False:* If all the observations of a histogram fall within the specification limits, then the process must be able to meet specifications.

_____ 4. *True or False:* If a lot plot has more than one peak, you might suspect that a mixture of observations from two different processes has been included.

END OF CHAPTER QUIZ

_____ 1. *True or False:* A frequency distribution is a graphical summary of a set of data.

_____ 2. *True or False:* A histogram and frequency distribution basically give the same information about a set of data.

_____ 3. *True of False:* Frequency distributions and histograms allow us to determine how well production can meet specifications.

_____ 4. *True or False:* Grouped frequency distributions are most useful for large amounts of data.

_____ 5. *True or False:* The number of cells that should be used in a grouped frequency distribution depends on the number of observations.

6. Suppose all observations in a set of data fall between 25 and 55. If a grouped frequency distribution with 5 cells is to be constructed, what should the cell width be?

7. For the following histogram, suppose the specifications are 145 to 160. What percentage of product is out of specification?

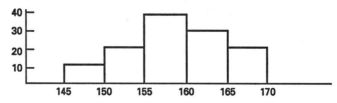

6

Process Capability

6-1. THE CONCEPT OF PROCESS CAPABILITY

Process capability is the range over which the natural variation of a process occurs, as determined by the system of common causes (natural variations in materials, machines and tools, methods, operators, and the environment). Process capability is measured by the proportion of output that can be produced within design specifications.

You were already introduced to the basic idea of process capability in the last chapter, when we discussed applications of histograms. If you recall, we discussed lot plots and their relationship with specifications. By comparing the distribution of process output to the specifications, we can see how well the process can produce output that conforms to these specifications.

There are three important elements of process capability: the design specifications, the centering of the process, and the range of variation. As we have seen before, the distribution of process output generally follows a bell-shaped curve. In this case, the quality characteristic in which we are interested is centered around some average value and spread out over some range.

Let us examine four possible situations that can occur if we compare the natural variation in a process with the design specifications.

1. In this situation, we see that the natural variation in the output is smaller than the tolerance specified in the design. You would expect that the process will almost always produce output that conforms to the specifications, as long as the process remains centered. Even slight changes in the centering or spread of the process will not affect its ability to meet specifications.

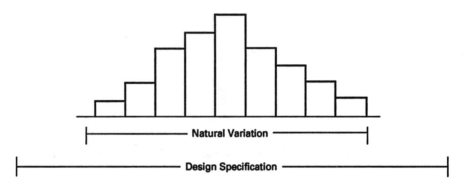

2. In this case, the natural variation and the design specification are about the same. A very small percentage of output might fall outside the specifications. The process should probably be closely monitored to make sure that the centering of the process does not drift and that the spread of variation does not increase.

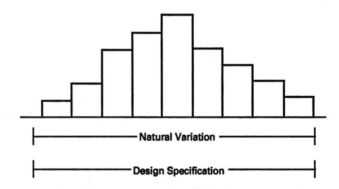

3. In this example, we see that it is impossible to meet the specifications all the time; the range of process variation is larger than the design specifications. The only way to improve product quality is to change the process—the materials, equipment, or work methods—to reduce the variation. Otherwise, the product must be carefully inspected in order to remove nonconforming items, but this adds unnecessary costs and wastes time.

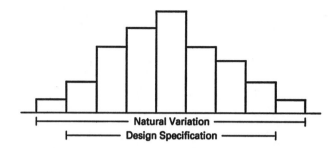

4. Finally, this example shows a situation in which the range of variation is smaller than the specifications, but the process is off-center. Some product will be produced that does not meet the lower specification. An adjustment in the centering of the process will result in nearly all output meeting specifications. Notice that the process is *capable* of meeting specifications, but cannot because of the centering problem.

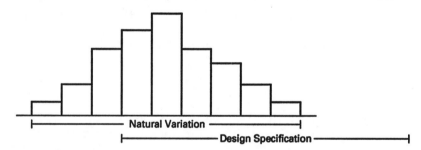

These examples show us why it is important to measure and to understand the variation in process output. If we can determine what the true state of quality is and how well a process can meet the design specifications, then we can take action to *improve* the process and the quality of our products. This is the purpose of process capability analysis.

The descriptive statistics that you have studied—frequency distributions, histograms, measures of location, and measures of variability—all help to provide quantitative information of the ability of a process to meet specifications. Statistical analysis will allow us to estimate approximately how much output will not meet specification, and to establish a measurement base for controlling a process.

Listed below is a sample of 120 measurements of steel shafts, accurate to .05 cm.

10.65	10.70	10.65	10.65	10.85
10.75	10.85	10.75	10.85	10.65
10.75	10.80	10.80	10.70	10.75
10.60	10.70	10.70	10.75	10.65
10.70	10.75	10.65	10.85	10.80
10.60	10.75	10.75	10.85	10.70
10.60	10.80	10.70	10.75	10.75
10.75	10.80	10.65	10.75	10.70
10.65	10.80	10.85	10.85	10.75
10.60	10.70	10.60	10.80	10.65
10.80	10.75	10.90	10.50	10.85
10.85	10.75	10.85	10.65	10.70
10.70	10.70	10.75	10.75	10.70
10.65	10.70	10.85	10.75	10.60
10.75	10.80	10.75	10.80	10.65
10.90	10.80	10.80	10.75	10.85
10.75	10.70	10.85	10.70	10.80
10.75	10.70	10.60	10.70	10.60
10.65	10.65	10.85	10.65	10.70
10.60	10.60	10.65	10.55	10.65
10.50	10.55	10.65	10.80	10.80
10.80	10.65	10.75	10.65	10.65
10.65	10.60	10.65	10.60	10.70
10.65	10.70	10.70	10.60	10.65

We may construct a frequency distribution and histogram for this data. This is a good practice problem for review. Take a moment to tally the observations, compute the mean and standard deviation, and draw a histogram. A worksheet for your calculations and space to draw a histogram are provided next.

x	Tally	f	x^2	fx^2	fx
10.50					
10.55					
10.60					
10.65					
10.70					
10.75					
10.80					
10.85					
10.90					

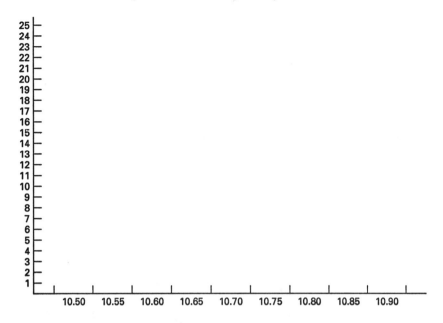

Your results should agree with those below.

x	f	x^2	fx^2	fx
10.50	2	110.2500	220.5	21.0
10.55	2	111.3025	222.605	21.1
10.60	13	112.36	1460.68	137.8
10.65	25	113.4225	2835.5625	266.25
10.70	22	114.49	2518.78	235.4
10.75	24	115.5625	2773.5	258.0
10.80	16	116.64	1866.24	172.8
10.85	14	117.7225	1648.115	151.9
10.90	2	118.81	237.62	21.8
	120		13783.6025	1286.05

$$\bar{x} = 1286.05/120 = 10.71708333$$

$$s = \sqrt{\frac{13783.6025 - 120(10.71708333)^2}{119}} = .0868437784$$

We will approximate $\bar{x}$ by 10.7171 and s by .0868.

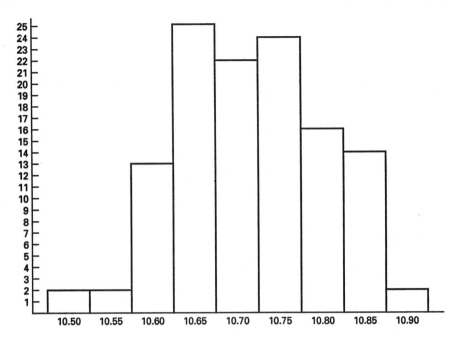

As we stated before, an important fact in statistics is that nearly all values of a frequency distribution fall within three standard deviations of the mean. Therefore, we can determine the natural variation of this process by adding and subtracting $3s$ to the mean $\bar{x}$.

$$\bar{x} + 3s = 10.7171 + 3(.0868) = 10.9775$$

$$\bar{x} - 3s = 10.7171 - 3(.0868) = 10.4567$$

This tells us that we can expect nearly all the measurements from this process to fall between 10.4567 and 10.9775 *as long as the process remains in control*. The process *must* have been in statistical control when the process capability study was performed. The range 10.4567–10.9775 is a measure of the natural variation in the process, or the process capability. These "capability limits" should never be confused with control limits.

Data from control charts can also be used to compute the standard deviation of a process and to establish a measure of process capability. In Chapter 4 we saw that an estimate of the standard deviation is

$$s = \bar{R}/d_2$$

where d_2 is a number that depends on the sample size used in the control chart.

Figure 6-1. Control and capability

For example, if the control chart was constructed using samples of size 5, then $d_2 = 2.326$ (see Table 4.2). All you need to do is divide the average range by this value to estimate the standard deviation. So if, for example, you have an average range $\bar{R} = .0315$, the standard deviation would be estimated as

$$s = \bar{R}/d_2 = .0315/2.326 = .0135$$

This method is not as accurate as calculating the standard deviation directly from the data, so we recommend it only if a quick estimate is needed.

We emphasize that process capability is not related to statistical control. Figure 6-1 shows four possible combinations of control and capability. We would like a process to be in control and capable of meeting specifications (box A in the figure). A process may be in control but not capable of meeting specifications (box B). This means that there is a problem due to common causes resulting in too large a variation. Management must act to improve the process to reduce the variation. Similarly, a process may be capable, but not be in control (box C). In this situation, special causes should be identified and eliminated. If a process is not capable *and* not in control (box D), steps should be taken *first* to get it in control, and then to attack the common causes that prevent a capable process.

REVIEW QUIZ 6-1

_____ **1.** *True or False:* Process capability is the tolerance that is specified on a blueprint.

_____ **2.** *True or False:* Process capability is measured by the proportion of output that can be produced within specifications.

_____ **3.** *True or False:* In the following situation, you would expect to have a high percentage of nonconforming output.

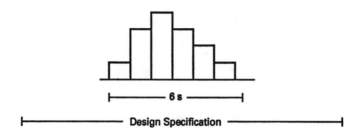

4. *True or False:* In order to measure process capability for some quality characteristic, you need to know the standard deviation of that quality characteristic.

5. A company that produces electric motors constructed $\bar{x}$ and R charts for an important part. The overall mean is 3.9376 and the average range is .00077. Samples of size 5 were used. What is an estimate of the standard deviation?
   a. .000154
   b. 1.69286
   c. .00077
   d. .00033

6-2. PROCESS CAPABILITY INDEX

The relationship between the natural variation of a process and the design specifications is often quantified by a measure called the *process capability index, Cp*. Many manufacturers use Cp to monitor the quality of their suppliers, and it is even used in purchasing contracts.

The process capability index is the ratio of the tolerance width to the natural process variation, or

$$Cp = \frac{\text{tolerance width}}{\text{natural variation}} = \frac{USL - LSL}{6s}$$

where USL and LSL are the upper and lower specification limits, respectively. Recall that all values of a distribution will usually fall within three standard deviations on either side of the mean. Thus, $6s$ is a good measure of the natural variation of a process. We discussed this in the previous section.

How do we interpret Cp? Suppose that the tolerance spread (USL − LSL) is 6, and the standard deviation of a process is 1.0, and the process is centered between the specification limits. Then

$$Cp = \frac{USL - LSL}{6s} = \frac{6}{6(1.0)} = 1.0$$

Figure 6-2 illustrates this situation. Since the tolerance spread is equal to the natural variation, the value of Cp is 1. In such situations, the process is just barely capable of producing within the specifications.

If the standard deviation is 2.0, then

$$Cp = \frac{USL - LSL}{6s} = \frac{6}{6(2.0)} = 0.5$$

A larger standard deviation means that the natural variation is greater; in this case, it is twice the tolerance spread. Whenever the natural variation

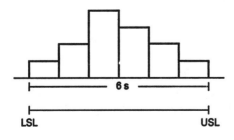

Figure 6-2. An example of $Cp = 1.0$

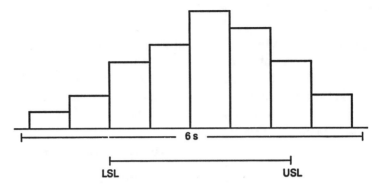

Figure 6-3.   An example of $Cp = 0.5$

is larger than the tolerance spread, Cp will be less than 1.0. This is illustrated in Figure 6-3. A large proportion of nonconforming product can be
expected in these situations.

As a final example, suppose that the standard deviation is 0.5. Then

$$Cp = \frac{USL - LSL}{6s} = \frac{6}{6(0.5)} = 2.0$$

The natural variation is smaller than the tolerance spread, as Figure 6-4
illustrates. We would expect very good quality in this situation. Whenever Cp is greater than 1.0, the process is capable of meeting specifications. The higher Cp is, the better. Many Japanese and American manufacturers now require Cp indexes in the 4 to 6 range.

The centering of the process is a very important assumption in using
the Cp index. One of the disadvantages of Cp is that it does not take
centering into account. For instance, each of the following examples has a
Cp of 1.0, but the center of each distribution is different.

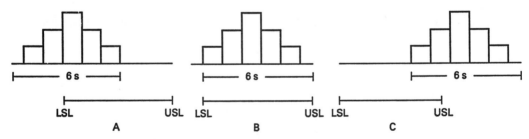

In A and C, we see that if the process is off-center, some nonconforming
product will be produced, even though the natural variation is the same as
the tolerance spread. These processes can probably be brought into conformance by adjusting the centering.

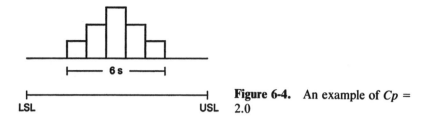

Figure 6-4.   An example of $Cp = 2.0$

The value of Cp alone will not tell you if the process is centered. To include information on process centering, one-sided process capability indexes are used. These are defined and calculated as follows.
Upper process capability index:

$$Cpu = \frac{USL - \bar{x}}{3s}$$

Lower process capability index:

$$Cpl = \frac{\bar{x} - LSL}{3s}$$

Let us see what these mean. Cpu is the ratio of the distance between the upper specification limit and the process mean (USL $- \bar{x}$) to half the natural variation ($3s$). Cpl is the ratio of the distance between the process mean and the lower specification limit ($\bar{x} -$ LSL) to half the natural variation ($3s$). If the process is centered between the specification limits, then $Cp = Cpu = Cpl$ since both the quantities in the ratio of Cpu and Cpl are exactly equal to half those in Cp. This is illustrated in Figure 6-5.

However, suppose that the process is not centered between the specification limits as shown in Figure 6-6. Since $\bar{x} -$ LSL is larger than USL $- \bar{x}$, the value of Cpl is larger than the value of Cpu. This means that the process is more capable of meeting the lower specification limit than it is of meeting the upper specification limit.

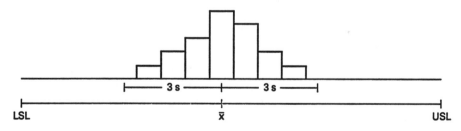

Figure 6-5.   A process centered within specifications

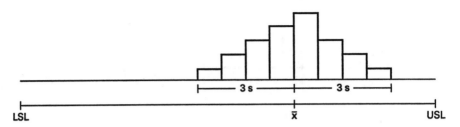

Figure 6-6. An off-center process.

We interpret the actual value of *Cpl* or *Cpu* in the same way as we do for *Cp*. Values greater than 1 mean that the process is capable of meeting that specification nearly 100 percent of the time, while values less than 1 are undesirable. One-sided limits are useful in applications in which only one specification limit is of interest. For example, in filling boxes of consumer products, federal laws require that the weight of each box be no less than a specified value. In such cases, the value of *Cpl* is important to monitor.

Often, the index *Cpk* is used, where *Cpk* is the *smaller* of *Cpu* and *Cpl*. *Cpk* defines the worst case between the upper and lower capability indexes. Let us use the following example. Suppose that USL = 6.00 and LSL = 5.00. A study determined that $\bar{x}$ = 5.60 and s = .15:

$$Cp = \frac{USL - LSL}{6s} = \frac{6.00 - 5.00}{6(.15)} = \frac{1.00}{.90} = 1.11$$

$$Cpu = \frac{USL - \bar{x}}{3s} = \frac{6.00 - 5.60}{3(.15)} = \frac{.40}{.45} = .89$$

$$Cpl = \frac{\bar{x} - LSL}{3s} = \frac{5.60 - 5.00}{3(.15)} = \frac{.60}{.45} = 1.33$$

$$Cpk = \text{smaller of .89 and 1.33} = .89$$

The value of *Cp* = 1.11 tells us that the process variation is small enough to meet specifications *if* the process is centered. However, the values of Cpu and Cpl tell us that the process is not centered and that we cannot expect the process to meet the upper specification limit very well. We would expect the process to meet the lower specification limit nearly all the time. The value of *Cpk* tells us that we cannot meet at least one of the specifications very well.

REVIEW QUIZ 6-2

_____ **1.** *True or False:* The process capability index Cp is the ratio of the design tolerance of a quality characteristic to its natural variation.

_____ **2.** *True or False:* A value of Cp greater than 1 means that the process capability is good, even if the process is not centered.

_____ **3.** *True or False:* Both Cpu and Cpl cannot be greater than 1.0 at the same time.

_____ **4.** *True or False:* Cpk is the average value of Cpl and Cpu.

END OF CHAPTER QUIZ

_____ **1.** *True or False:* Process capability is measured by comparing the design specifications to the centering of the process.

_____ **2.** *True or False:* If the design specification is larger than the natural variation, then quality problems will result.

_____ **3.** *True or False:* The natural variation of a process can be estimated by computing the mean plus or minus three standard deviations.

Use the following data to answer the questions below.

Upper specification limit = 10.50

Lower specification limit = 10.38

Actual process average = 10.47

Process standard deviation =   .04

4. Compute the value of Cp. Is the process capable of meeting specifications?

5. Compute the value of Cpu. What does this mean?

6. Compute the value of *Cpl*. What does this mean?

7. Find *Cpk*.

7

Implementing SPC and Other Tools for Quality Improvement

7-1. IMPLEMENTING STATISTICAL PROCESS CONTROL

To this point we have focused discussion on three basic tools of statistical process control: tally sheets for collecting and recording data, control charts for monitoring processes over time, and frequency distributions and histograms to determine the capability of a process to meet specifications. These tools are simple to use. However, putting them to work on the production floor to improve quality requires much more work and commitment.

People experienced in using SPC have found that four "key success factors" are necessary to make SPC work. First, SPC requires *top management commitment*. The top managers in a company must understand the importance of quality and commit the resources necessary to get quality improvement projects going. This includes the financial resources for proper measurement instruments, calculators, perhaps computers and software, and training employees for learning the mechanics of SPC. Using SPC properly disrupts production to some extent. It costs time and money. Management must demonstrate that this is not a "fad" that will disappear in a few months, but an ongoing commitment to improve the product and to help the work force do a better job.

Second, successful SPC projects need a *champion,* that is, some individual in the company who has both the responsibility and the authority to make it successful—someone who goes to work every day with only this task on his or her mind. Any kind of business project invariably fails if there is no champion to promote it and ensure its success.

Third, *tackle only one problem at a time.* If your company has never used SPC, it does not make sense to try to set up control charts throughout an entire plant or even within a whole department all at once. Mistakes will be made at the beginning. Using SPC effectively is a learning process. Pick a process that stands to benefit the most from SPC, and one that will have high visibility both to top management and to other workers and departments. It only takes one good success story to get others to want to copy it! Successful SPC projects need to be publicized throughout the company.

Finally, *education and training of all employees* is absolutely necessary. Everyone needs to understand why quality is important, what SPC can do to improve quality, and how it works. Workers must understand that SPC will benefit them and make their job easier; it is not a scheme set up by management to place blame on workers.

Implementing an SPC project is not difficult. As we have said, you should carefully select the first process for which to install SPC. You must *understand* the process well. You will need to be able to identify assignable causes that can cause the process to be out of control. This cannot be done for a process that is not well understood. Operators, supervisors, quality technicians, and engineers should all be involved in the project.

Before collecting any control chart data, you must determine how well your measurement system is able to accurately and precisely (these terms were discussed in Chapter 1) measure the quality characteristics that will be monitored with SPC. If either the accuracy or precision of the gaging is not acceptable, it makes no sense to go further. Bad measurements will only produce bad results, or, as the computer scientists say, "garbage in, garbage out."

If gaging is acceptable, you can proceed to collect initial data to construct a control chart, using the procedures described in Chapter 3. Use 25 to 30 samples to begin with, plot the sample points on the appropriate charts, and compute the control limits. Analyze the charts for any special causes. If special causes are present, track them down and make corrections to the process. Repeat this step if necessary to find the correct control limits before using the charts as an ongoing monitoring tool.

Only *after* the process has been brought into control can you perform process capability analysis. Process capability describes the system of common causes of variation. To improve quality, you must keep the special causes out of the process and try to discover how to reduce the variation due to common causes.

Problem solving requires you to be able to

1. identify problems to solve
2. determine the causes underlying the problems
3. collect data to verify the source of problems
4. take correct action.

Several tools are available that can help you to do this. We shall discuss these in the remainder of this chapter.

REVIEW QUIZ 7-1

_____ 1. *True or False:* SPC can always be implemented successfully by quality control departments without top management support.

_____ 2. *True or False:* In using SPC, a company should begin with small projects that are likely to show good benefits from SPC.

_____ 3. *True or False:* A process capability study should always come before using control charts.

7-2. PARETO ANALYSIS

Pareto analysis is a technique highly advocated by Dr. Joseph Juran to identify the most critical quality problems. This technique is named after an Italian economist, Vilfredo Pareto, who determined that 85 percent of the world's wealth was owned by 15 percent of the people. In terms of quality, we can usually say that a large percentage of the problems are determined by a small percentage of the causes. Identifying these "vital few" causes and correcting them will give us the most improvement for our efforts. We should not worry about the "trivial many" causes that account for only a few quality problems at this time.

To conduct a Pareto analysis, we collect data on the frequency of different causes, or types of quality problems. Then we *sort* these problems in order of frequency and calculate the percentage of each as well as the cumulative percent. Often, we also express the result using a graphical diagram called a Pareto diagram. This method is best seen using an example.

Suppose we have recorded the number of occurrences of different circuit board defects over the past month:

Defect type	Frequency
short circuit	32
missing component	9
solder	87
improper wiring	12
total	140

The first step in Pareto analysis is to sort the defects by frequency, largest first:

Defect type	Frequency
solder	87
short circuit	32
improper wiring	12
missing component	9
total	140

Next, we compute the percentage of each defect type.

What is the percentage of solder defects?

Since there are 140 total defects and 87 solder defects, the percentage of solder defects is $87/140 = .6214$ or 62.14 percent. If we do this for each of the defects, we have:

Defect type	Frequency	Percentage
solder	87	62.14
short circuit	32	22.86
improper wiring	12	8.57
missing component	9	6.43
total	140	100.00

Finally, we compute the *cumulative percentage* for each defect type:

Defect type	Frequency	Percentage	Cumulative percentage
solder	87	62.14	62.14
short circuit	32	22.86	85.00
improper wiring	12	8.57	93.57
missing component	9	6.43	100.00
total	140	100.00	

We see that 85 percent of the defects are accounted for by the first two defect types, and that over 60 percent of the defects are due to soldering problems alone. This tells us that we will be better off attacking the reasons for solder defects before working on any other problem.

This information is often summarized in a Pareto diagram like the one shown in Figure 7-1. A Pareto diagram is like a histogram. The vertical

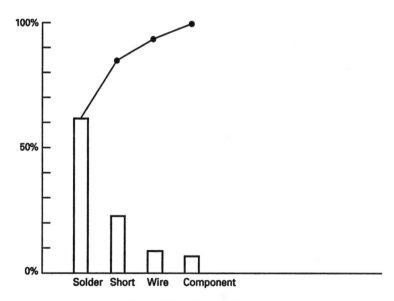

Figure 7-1. Pareto diagram

bars correspond to the percentage of each defect. The line above shows the cumulative percentages. The diagram shows employees very clearly which problems are most important.

Pareto analysis gives factual evidence on which to base quality improvement projects. It helps workers and managers to reach agreement on which problems to attack first. Finally, it can be used as a follow-up tool to show the results that actually are achieved.

REVIEW QUIZ 7-2

Suppose we have collected the following data on defects on auto body panels:

Defect type	Frequency
surface scratch	13
dent	25
trim misalignment	3
paint smear	9

1. Sort the defects in order of frequency.
2. Calculate the percentage of each.
3. Calculate the cumulative percentages.
4. Draw the Pareto diagram.

7-3. CAUSE AND EFFECT DIAGRAMS

When a process is discovered to be out of control, what action should be taken? When other quality problems are discovered, how do you determine what causes them? We cannot give a definitive answer to these questions, because it depends on the type of process or problem. Control charts only give a *statistical signal* that something is wrong. Likewise, Pareto analysis also provides a signal. Neither can tell you what the problems are. Only the expert knowledge of the production operators, supervisors, engineers, and quality technicians can get to the source, or root cause, of the problem.

A simple tool exists, however, to help you "map out" this knowledge to determine the causes of quality problems. This tool, called a *cause and effect diagram,* was developed by Dr. K. Ishikawa in Japan. Sometimes it is called an Ishikawa diagram, or for reasons that will be obvious in a moment, a fishbone diagram. The general structure of a cause and effect

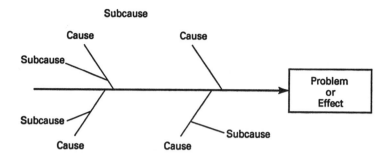

Figure 7-2.   General structure of a cause and effect diagram

diagram is shown in Figure 7-2. At the end of the main stem is listed the effect, or problem that we wish to analyze. The "fishbones" that branch out from the main stem are the major causes that can be identified. The major causes will usually be operators, materials, machines, methods, or measurement. From each of these, we can draw branches representing subcauses in each of these categories.

The general steps in constructing a cause and effect diagram are:

1. Define the problem, or effect, that you wish to diagnose and solve. Write it down at the end of the main stem of the cause and effect diagram.

2. Determine the major factors that you believe might cause the problem and write them down next to branches coming out from the main stem.

3. For each main factor, define and list any subfactors that may contribute to the problem.

Cause and effect diagrams are most useful in group problem-solving situations with a leader to guide the "brainstorming" process. In themselves, they do not provide any answers. However, the process by which they are constructed is enlightening and promotes better understanding of the problem by all concerned. The cause and effect diagram provides a basis for further discussion and diagnosis of the problem and helps in reaching conclusions faster than unstructured approaches.

To illustrate a simple example, suppose that a company is receiving many complaints from its customers about its products being damaged. The problem to be studied is "damaged product." In a group problem-solving session, it might be suggested that major causes of this problem are transportation, packaging material, packaging method, and unpacking pro-

Figure 7-3. Example of a cause and effect diagram

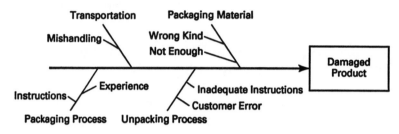

Figure 7-4. Example of a cause and effect diagram

cess. These would be drawn on the cause and effect diagram as shown in Figure 7-3.

Further discussion might hone in on the individual causes and suggest possible subcauses, such as inadequate packaging instructions or amount of work experience for the packaging process. These would be listed on the diagram as in Figure 7-4. The discussion would continue until the group feels that all possible avenues have been exhausted.

When a cause and effect diagram is constructed, we can see at a glance the potential causes and relationships. We can now focus attention on the most likely ideas for further study. The next logical step is to collect some data to verify whether there really is a cause and effect relationship.

REVIEW QUIZ 7-3

1. *True or False:* Cause and effect diagrams clearly tell which cause is the most likely.
2. Cause and effect diagrams are also called:
   a. Ishikawa diagrams
   b. fishbone diagrams
   c. all of the above

3. *True or False:* Cause and effect diagrams are best used in group problem-solving sessions.

7-4. SCATTER DIAGRAMS

Scatter diagrams show the relationship between paired data and are useful in helping to understand whether there really is a cause and effect relationship between factors. Therefore, they are often used to attempt to verify cause and effect relationships suggested by Ishikawa diagrams. For example, suppose that work experience of the packaging employees is thought to be the most likely factor in the previous example. We would need to collect data on the damaged products and the employee who packed them, and then record the amount of experience the employees have. We might find the following for the past month:

Employee	Number of damaged packages	Months of work experience
A	2	18
B	8	4
C	1	24
D	11	3
E	9	6
F	4	10

To construct a scatter diagram, you should collect *paired* samples of data that you suspect are related. Draw the horizontal and vertical axes of a graph and scale the axes to the range of the data. Finally, plot the data on the graph. If more than one point falls at the same place, make circles around that point to note this. For this simple example, we would plot the months of work experience on the horizontal axis and the number of damaged packages on the vertical axis as shown in Figure 7-5.

From Figure 7-5 it appears that the number of damaged packages is related to the amount of work experience. The more experienced workers have a lower rate of damaged product. This suggests that perhaps the packers need improved training and instruction on technique.

Scatter diagrams can tell us a lot about a process. The more data points we have, the better will be our conclusions. Usually, about 50 to 100 points are adequate (the example above used only a small number just to illustrate the idea). Figure 7-6 shows several examples of scatter diagrams. In (a), we see a *positive relationship* between the two variables; as one gets larger, so does the other. In (b), we see the opposite; as the variable on the horizontal axis gets larger, the variable on the vertical axis

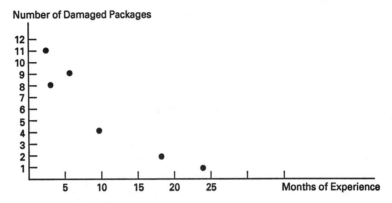

Figure 7-5. Example of a scatter diagram

gets smaller. Figure 7-6(c) shows no relationship at all between the variables.

We may use scatter diagrams to illustrate the results of *controlled experimentation*. In a controlled experiment, we vary one or more factors to determine what effect they have on some response. For example, in a chemical process, we may vary temperature, pressure, and amount of additive to determine the effect on yield of the final product. Controlled experimentation is a powerful method to improve quality. It does, however, require more advanced statistical methods than we have studied in this book. Therefore, we recommend that it only be conducted under the guidance of a competent statistician.

Figure 7-6. Examples of scatter diagrams

REVIEW QUIZ 7-4

_____ **1.** *True or False:* Scatter diagrams show the trend of a variable over time.

2. Suppose the following scatter diagram shows the relationship between the number of production errors and the amount of overtime of employees.

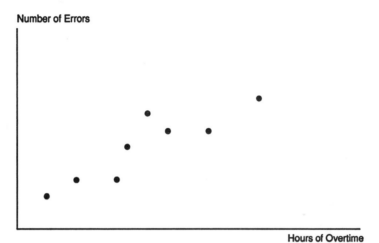

Number of Errors

Hours of Overtime

This diagram suggests:
a. as overtime increases, the number of errors decreases
b. as overtime increases, the number of errors increases
c. no relationship
d. errors cause overtime

3. What does the following scatter diagram suggest?

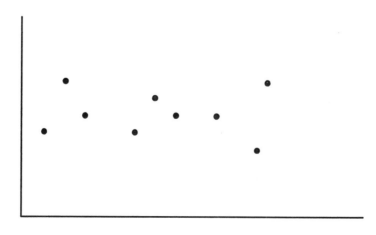

    a. positive relationship
    b. negative relationship
    c. no relationship

END OF CHAPTER QUIZ

1. *True or False:* It is best to install SPC throughout a company at the beginning of a quality improvement program.
2. *True or False:* A process must be well understood in order to use SPC effectively.
3. *True or False:* The adequacy of the measurement instruments is not important for SPC.
4. Which technique is most appropriate to determine why a trend has been found in a control chart?
   a. Pareto analysis
   b. Cause and effect diagram
   c. Scatter diagram
5. Which technique should be used to determine the quality problem whose solution will result in the largest savings?
   a. Pareto analysis
   b. Cause and effect diagram
   c. Scatter diagram
6. *True or False:* Studies using scatter diagrams are usually done after cause and effect relationships are suggested.
7. *True or False:* You must collect a lot of data to use cause and effect diagrams.

A

Basic Math Skills

A-1. MATH SKILLS ANALYSIS

Being able to perform some basic mathematical calculations involving whole numbers, fractions, and decimals is important to successfully learn how to use and apply statistical process control. In this section, we will ask you to solve a set of problems dealing with arithmetic, fractions, and decimals. You will probably be using a calculator for doing SPC calculations. If you wish to use one now, by all means do so! After checking your answers with those in appendix B, you may feel that you need additional review and practice. If so, we strongly suggest that you study the remaining sections of this chapter before continuing with your study of SPC.

PART 1: BASIC ARITHMETIC

Solve the following problems involving addition, subtraction, multiplication, or division.

1.   18
          49
        +12

2.   33.622
          47.721
          21.276
        +68.173

3.   17.837
        −12.527

4.   63.721
        −   .312

5.   23
        ×74

6.   4,963
        ×1,579

7. 20)‾2040

8. 251)‾157,628

PART 2: FRACTIONS

Simplify the following fractions. Write your answer in the form of x/y (for example, 3/5).

1. 4/8 =

2. 20/32 =

3. 15/125 =

In the next two problems, find the numerator for the second fraction so that both represent the same number.

4. 4/5 = ?/20

5. 16/100 = ?/25

Solve the following problems. Simplify if possible.

6. 1/9 + 3/9 =

7. 3/5 + 3/4 =

8. 5/7 − 2/7 =

9. 7/8 − 7/9 =

10. 2/6 × 3/5 =

11. $6 \times 5/8 =$

12. $(3/8)/(1/2) =$

13. $(10/16)/8 =$

PART 3: DECIMALS

1. What is the decimal number for nine-tenths?
2. What is the decimal number for five and four-hundredths?
3. What is the decimal number for three and seventy-seven hundred thousandths?
4. What is the decimal number for 6/10?
5. What is the decimal number for 7 16/10000?
6. Round .3876 to the nearest thousandth.
7. Round 3.7695 to the nearest hundredth.
8. Round .07514 to the nearest ten thousandth.

Solve the following problems.

9.   4.072   **10.**   9.07
  +16   −2.79

11.   111.03   **12.**   .396
  − 6.4   × 16

13.   7.69   **14.** 19)8.7628
  × 3.5

15. .35)7763

16. What is the square of 2?
17. What is the square of 24?
18. What is the square of 5.16?
19. What is the square root of 9?
20. What is the square root of .16?
21. What is the square root of 2.25?

END OF MATH SKILLS ANALYSIS

A-2. MATHEMATICS REVIEW: FRACTIONS

A fraction is a part of a whole. For example, suppose we have the eight shapes shown in Figure A-1. The fraction, or proportion, of squares in this group is 3/8. Likewise, the fraction of circles is 3/8, and the fraction of triangles is 2/8. In general, to express the fraction of things that possess some characteristic, we use the following formula:

$$\frac{\text{number of things that possess the characteristic}}{\text{total number of things}}$$

For example, suppose you inspect 500 parts and find that 3 are defective. The *fraction defective* is

$$\frac{\text{number defective}}{\text{total number inspected}} = \frac{3}{500}$$

What fraction of the shapes shown in Figure A-1 are boxes *or* circles?

Since there are three boxes and three circles, they comprise six of the eight total shapes. The correct answer is six-eighths, or three-fourths.

Figure A-1.   A set of eight shapes

When dealing with fractions, we refer to the number in the top part of the fraction as the *numerator*. Likewise, the number in the lower part of the fraction is referred to as the *denominator*. Thus, in the fraction 6/8, 6 is the numerator and 8 is the denominator.

In the fraction 6/8 it is easy to see that both 6 and 8 can be divided evenly by 2. Dividing both the numerator and denominator by 2 gives the reduced fraction 3/4. The fraction 3/4 cannot be reduced further because the only number that can be divided evenly into both the numerator and the denominator is 1.

Sometimes a fraction that is reduced once can be reduced one or more additional times. For example, suppose that we have the fraction 8/16. Clearly, one number that can be divided into both 8 and 16 is 2. If we do this, the fraction becomes 4/8. However, the fraction 4/8 can be reduced further because both 4 and 8 can be evenly divided by 2 (or by 4). To fully reduce the fraction 4/8, we should divide both 4 and 8 by the largest possible number that we can. If we divide both the numerator and the denominator by 4, we reduce the fraction to 1/2. We cannot reduce 1/2 any further.

There often are several different ways of reducing a fraction. For example, for the fraction 8/16 we could observe that both 8 and 16 are evenly divisible by 8. Dividing both the numerator and denominator by 8 gives 1/2. In this case we were able to fully reduce the fraction in one step.

REVIEW QUIZ A-2

1. What is the fraction of boxes in the group of shapes below?

2. The number in the top portion of a fraction is called the _____ .
   a. numerator
   b. denominator
3. The number in the lower portion of a fraction is called the _____ .
   a. numerator
   b. denominator
4. What is the fully reduced fractional form of 2/3?

5. What is the fully reduced fractional form of 9/12?

6. What is the fully reduced fractional form of 14/35?

A-3. MATHEMATICS REVIEW: DECIMAL NUMBERS

Decimal numbers are widely used in shop applications and in quality control. A decimal number is one that has digits written to the right of the decimal point (.). The decimal part represents a fraction. For example, we often use decimal numbers in part specifications. Suppose that we desire to drill 10-mm holes in a part. Let us assume that the hole diameters of $9\frac{1}{2}$ mm to $10\frac{1}{2}$ mm are acceptable. We can represent the specification for the hole diameter as a decimal number as follows:

$$\text{hole diameter} = 10.0 \pm 0.5 \text{ mm}$$

The decimal number 0.5 represents the amount that the hole diameters are allowed to vary. In this situation the drilled holes may vary by 0.5 (or one-half) mm.

We can also represent the interval of acceptable hole diameters using decimals as follows:

$$\text{hole diameter} = 9.5 \text{ to } 10.5 \text{ mm}$$

In this book we will only go as far as four decimal places.

Suppose that A, B, C, and D are single digits in the following decimal number:

$$X.ABCD$$

X represents a whole number such as 1, 2, 10, 100, 2433, and so on. A represents the tenths place. For example, the fraction 3/10 is written in decimal form as 0.3 and is called three-tenths. B represents the hundredths place. The fraction 75/100 is written in decimal form as 0.75 and is called seventy-five hundredths. C represents the thousandths place. The fraction 22/1000 is written in decimal form as 0.022 and is called twenty-two one-thousandths. Finally, D represents the ten-thousandths place. The fraction 25/10000 is written as 0.0025 and is called twenty-five ten-thousandths.

As the previous examples illustrate, fractions can be easily converted into decimal numbers. To convert a fraction to a decimal, we simply divide the numerator by the denominator of the fraction. For example, consider the fraction 1/2. To convert this to a decimal, we divide 1 by 2. This gives us the decimal 0.5. As another example, consider the fraction 3/4. To convert this to a decimal, divide 3 by 4. This gives the number 0.75. As a final example, consider the fraction 1/3. When we divide 1 by 3, the result is an endless string of 3s to the right of the decimal point. When this occurs, we round the fraction to a fixed number of decimal places. This depends on the precision that we require.

In quality control applications, we rarely require a precision of greater than four digits to the right of the decimal point. Thus, if we wanted to round the fraction 1/3 to four decimal places, we would get 0.3333.

To round any number to "n" decimal places, follow the procedure given below.

1. Write the first $n - 1$ digits exactly as they appear.
2. If the number after the nth digit is 5, 6, 7, 8, or 9, add one to the nth digit. If the number after the nth digit is 0, 1, 2, 3, or 4, write the nth digit exactly as it appears.

As an example, consider the decimal number 0.2456. If we round this number to one decimal place, we get 0.2, since the second digit is 4. If we round this to two digits, we get 0.25, since the third digit is 5. Finally, rounding this to three decimal places gives 0.246, since the fourth digit is 6.

One cautionary note. Precision and accuracy are two entirely different things. For example, if we say that the average car lasts 100.3456723 years, that would be a very precise statement. However, it would not be very accurate! Likewise, if we say that the average car lasts between 5 and 10 years, we would be rather accurate, but not very precise.

REVIEW QUIZ A-3

1. What is the decimal equivalent of 1/5?

2. What is the decimal equivalent of 2/3 rounded to three decimal places?

3. What is the decimal equivalent of 4/7 rounded to two decimal places?

A-4. MATHEMATICS REVIEW: BASIC OPERATIONS WITH DECIMALS

In this section we will review how to add and subtract decimal numbers. Adding and subtracting decimal numbers is very easy. The only thing you must remember is to keep the decimal points lined up. For example, suppose we would like to add 1.23 and 5.63 together.

To carry out this addition, we first arrange the two decimal numbers as follows:

$$\begin{array}{r} 1.23 \\ \underline{5.63} \end{array}$$

Notice that the decimal points line up in the same column. Once we have lined up the decimal points, we simply add (or subtract) the numbers beginning with the right-most column. We would obtain:

$$\begin{array}{r} 1.23 \\ \underline{5.63} \\ 6.86 \end{array}$$

We can add or subtract decimal numbers that have different numbers of digits to the right of the decimal point. We simply treat the "missing" digits as zeros. For example, suppose we want to add 1.2 and 3.456 together. First we line up the decimal points, add the "missing" zeros, and then add:

$$\begin{array}{r} 1.200 \\ \underline{3.456} \\ 4.656 \end{array}$$

To illustrate the process of subtracting decimal numbers, consider the following example:

$$\begin{array}{r} 9.7 \\ \underline{-5.432} \end{array}$$

We add the "missing" zeros so that both numbers have the same number of digits to the right of the decimal point, then perform the subtraction.

$$
\begin{array}{r}
9.700 \\
-5.432 \\
\hline
4.268
\end{array}
$$

The last mathematical concepts that we shall review are the square of a number and the square root of a number. If x is any number, then the square of x is written x^2 (called x squared) and is simply the value of x multiplied by itself. For example, $2^2 = 2 \times 2 = 4$; $5^2 = 5 \times 5 = 25$.

The opposite of squaring a number is taking its square root. For example, since 4 is the square of 2, 2 is the square root of 4; 25 is the square of 5, so 5 is the square root of 25. The square root of a positive number x is denoted by:

$$
\sqrt{x}
$$

For instance, $\sqrt{4} = 2$ and $\sqrt{25} = 5$. Square roots are difficult to compute by hand. Most of the time you must use a calculator. For statistical process control applications, you should have a calculator that finds square roots.

REVIEW QUIZ A-4

1. What is the sum of 4.323 and 1.56?

2. What is $5.6 - 3.25$?

3. What is the square of 3?

4. Find the square root of 16.

B

Answers to Review and End of Chapter Questions

CHAPTER 1

Review Quiz 1-1

1. *True*. Interchangeable parts necessitated close conformance to manufacturing specifications and made inspection an important activity. This led to the creation of separate quality departments.
2. *False*. "Fitness for use" means that the product meets the needs of the customer.
3. *True*.
4. *False*. SPC uses only simple mathematics.

Review Quiz 1-2

1. *True*. Common causes of variation consist of all the minor variations in a production process that cannot be fully understood, predicted, or controlled individually. Taken together, we can measure them in aggregate. Special causes *can* be identified.

2. *False*. Common causes are usually beyond the control of the operator but are the responsibility of management. Special causes can be identified by operators and usually corrective action can be taken by them.
3. *False*. Inspection alone is wasteful and uneconomical. Doing it right the first time should be the principal goal of production. Furthermore, many products cannot be inspected without ruining them.
4. *True*.
5. *False*. SPC can indeed lower costs and increase productivity by reducing the nonproductive time spent on rework, scrap, and so forth.

Review Quiz 1-3

1. *False*. Data that come from counting are called *attributes data*.
2. *False*. The collection of all possible items of interest in a particular study is called a *population*. A *sample* is a portion of all possible items of interest.
3. *True*.
4. *False*. If problems are identified, it is up to operators and supervisors to determine the causes and to make corrections. SPC alone cannot provide this information; it can only signal that a problem exists.
5. *True*. Weight can be measured on a continuous scale and therefore is an example of a variables measurement.
6. *False*. Accuracy refers to the ability to measure the true value; *precision* refers to repeated measurements.

Review Quiz 1-4

1.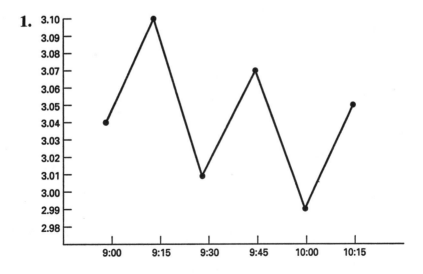

End of Chapter 1 Quiz

1. *False*. Variation is common in every production process.
2. *False*. Inspection alone will never help you to control and improve quality.
3. *True*.
4. The correct answer is (b). Variables data correspond to measurements along a continuous scale.
5. The correct answer is (a). Attributes data come from counting, not measurement.
6. *False*. A sample is a group of items selected from a population.
7. *True*.
8. *True*. SPC provides a signal that there might be a problem, but it is up to the operators, supervisors, and managers to identify the cause of the problem and correct it. Without SPC, however, you might not know that a problem exists.

CHAPTER 2

Review Quiz 2-1

1. (a) mean
2. (b) median
3. (c) mode
4. a. 147
   b. 146
   c. 144

Review Quiz 2-2

1. (a) range
2. (b) standard deviation
3. (b) sum
4. We usually use small letter s to denote the standard deviation of sample data.
5. The largest value is 9; the smallest value is 1. Therefore, the range is $9 - 1 = 8$.

6.

x	$x - \bar{x}$	$(x - \bar{x})^2$
4	.8	.64
3	−1.8	3.24
1	−3.8	14.44
7	2.2	4.84
9	4.2	17.64
		40.80

$$s = \sqrt{40.80/4} = \sqrt{10.2} = 3.194$$

End of Chapter 2 Quiz

1. *False*. Measures of location describe the centering of the data.
2. *True*.
3. *True*.
4. *False*. The mode usually is not as good a measure of central location as the median or mode. However, for large sets of data that are reasonably bell-shaped, the mode does provide a quick estimate of the central value of the data.
5. *True*.
6. *True*.
7. *False*. The range uses only the largest and smallest values.
8. *True*.
9. *False*. The smaller the standard deviation, the less variability there is in the data.

CHAPTER 3

Review Quiz 3-1

1. **(a)** in control
2. **(b)** out of control
3. *False*. Control charts are used to determine if special causes of variation are present.
4. *True*.
5. *False*. We record small samples at regular time intervals.
6. *True*.
7. *False*. The overall mean is the average of the sample means.

Review Quiz 3-2

1. *False*. We plot sample means, not individual observations.
2. *True*.
3. **(b)** *R*-chart
4. **a.** $UCL = D_4\overline{R} = 2.574(.84) = 2.16216$.
   b. $LCL = D_3\overline{R} = 0(.84) = 0$
   c. $UCL = \overline{\overline{x}} + A_2\overline{R} = 2.4 + 1.023(.84) = 3.25932$
   d. $UCL = \overline{\overline{x}} - A_2\overline{R} = 2.4 - 1.023(.84) = 1.54068$
5. *False*. Not necessarily. A single point outside the control limit may be a chance occurrence, although it is highly unlikely.
6. *True*.
7. $C - B - D - A - E$

Review Quiz 3-3

1. *False*. The sample size must be the same as the one used to construct the chart.
2. *True*.
3. *True*. Probably. However, we caution you that a point may fall outside a control limit even if the process is in control, but this is not very likely.
4. *True*.
5. *False*. There is no relationship between control limits and specification limits. Control limits depend on the variation in the process; specification limits are arbitrarily set by designers.

Review Quiz 3-4

1. *False*. There are a variety of other reasons why we might decide a process is no longer in control even though all the points fall within the control limits. For example, there might be trends or nonrandom patterns in the data.
2. *True*. We would expect about 50 percent of the points to fall either above or below the center line.
3. *False*. Unnatural patterns in the $\overline{x}$-chart can be caused by assignable causes in the R-chart. We should always make sure that the R-chart is in control before we analyze the $\overline{x}$-chart.
4. *Out of control*. Too many points appear above the center line.

5. *Out of control.* The data exhibit an upward trend.

6. *In control.*

7. *Out of control.* The data appear to "hug" the center line too closely.

8. *Out of control.* The data show wild fluctuations with points outside of the control limits.

9. *Out of control.* Seven of the last eight points are above the center line, indicating that a shift may have occurred.

Review Quiz 3-5

1. (a) trend

2. (b) single point beyond control limits

3. (d) sudden shift of process average

4. (e) cycles

5. (b) hugging the center line

6. (c) instability

7. (b) mixture

8. *False.* A shift down in the process average in the *R*-chart means that the process has become *more* uniform.

9. *True.*

Review Quiz 3-6

1. *True.*

2. *False.* Attributes data assume only two values such as good or bad and therefore are not measured.

3. *True.*

4. *True.*

5. $\bar{p} = (0 + .1 + .2 + 0 + .2)/5 = 0.1.$

6. $s = \sqrt{\dfrac{.2(1 - .2)}{10}} = .1265$

End of Chapter 3 Quiz

1. *False.* If only common causes are present, the process is said to be in control.

2. *True.*

3. *False*. At least 25 samples should be used for the purpose of constructing a control chart.
4. mean = $(5 + 7 + 10 + 4 + 4)/5 = 30/5 = 6.0$
5. range = $10 - 4 = 6$
6. *False*. The overall mean is the average of the sample means.
7. *False*. Control limits for the R-chart should be computed first and you should make sure that it is in control.
8. *False*. The sample size must be the same.
9. *True*. This *may* happen for a process in control, but it is highly unlikely.
10. *True*.
11. *True*.
12. *False*. A trend is a sequence of points that move toward one of the control limits. *Cycles* have alternate peaks and valleys.
13. *False*. You would expect changes from a new group of operators to be sudden, not gradual. This would more likely cause a shift.
14. *True*. This should be checked first.
15. *True*.
16. *False*. $\bar{x}$- and R-charts apply only to variables data.
17. *False*. Defect refers to a single nonconforming quality characteristic; defective refers to an item having one or more defects.
18. *True*.

CHAPTER 4

Review Quiz 4-1

1. *True*.
2. *False*. Different factors must be used to compute the proper control limits for each type of chart.
3. *False*. Some points that are within control limits on one chart may fall out of control on the other.

Review Quiz 4-2

1. *False*. The formulas do not apply since for samples of size 1, there is no variability.

2. *True*.

3. *True*.

4. *True*.

Review Quiz 4-3

1. *False*. They are used to track the number of defects, not defectives.

2. *True*. The sample size is not constant.

3. *True*.

End of Chapter 4 Quiz

1. (b) *s*-chart

2. *True*.

3. *False*. We must use a moving range chart.

4. *False*. Defect refers to a single nonconforming quality characteristic; defective refers to an item having one or more defects.

5. The overall mean is 2.122; the average standard deviation is 0.111. Control limits are given by:

$\bar{x}$-chart:

$$UCL = 2.122 + 1.099(.111) = 2.244$$

$$LCL = 2.122 - 1.099(.111) = 2.000$$

s-chart:

$$UCL = 1.815(.111) = .201$$

$$LCL = .185(.111) = .021$$

6. (a) Two-period moving range: 0.5, 1.1, 3.1, 1.2, 1.8, 0.7, 0.0, 1.1, 1.0, 1.7, 1.4, 5.9, 4.6, 0.1, 0.5, 1.0, 0.8, 0.1, 0.4, 1.0, 0.4

$$\bar{R} = 1.37; \bar{x} = 11.57$$

Control limits on *x*-chart:

$$UCL = 11.57 + 3(1.37)/1.128 = 15.21$$

$$LCL = 11.57 - 3(1.37)/1.128 = 7.93$$

For moving range chart:

$$UCL = 3.267(1.37) = 4.48$$

$$LCL = 0$$

(b) Three-period moving range: 1.1, 3.1, 3.1, 1.8, 1.8, 0.7, 1.4, 1.4, 2.7, 1.7, 5.9, 5.9, 4.7, 0.5, 1.0, 1.8, 0.9, 0.4, 1.4, 1.0

$$\bar{R} = 2.12; \bar{x} = 11.57$$

Control limits on x-chart:

$$UCL = 11.57 + 3(2.12)/1.693 = 15.33$$

$$LCL = 11.57 - 3(2.12)/1.693 = 7.81$$

For moving range chart:

$$UCL = 2.574(2.12) = 5.46$$

$$LCL = 0$$

7. Total number defective = 176; number of samples = 10

$$\bar{c} = 176/10 = 17.6$$

$$UCL = 17.6 + 3\sqrt{17.6} = 30.19$$

$$LCL = 17.6 - 3\sqrt{17.6} = 5.01$$

CHAPTER 5

Review Quiz 5-1

1. The correct answer is (b). A frequency distribution summarizes a set of data by showing the frequency of a particular value or observation.

2. a. The smallest observation is 1.324
   b. The largest observation is 1.328
   c.

Observation	Frequency
1.324	1
1.325	4
1.326	5
1.327	3
1.328	2

   d. The most frequent observation is 1.326.
   e. The least frequent observation is 1.324.

Review Quiz 5-2

1. A graphical representation of a frequency distribution is called a *histogram*.

2. The values of the observations in a simple frequency distribution or the cell limits in a grouped frequency distribution are placed on the *horizontal* axis.

3. The frequencies of the observations or cells are placed on the *vertical* axis.

4. The correct answer is (c). Remember that there should be no gaps in the scale on the horizontal axis.

Review Quiz 5-3

1.

x	f	x^2	fx^2	fx
.5	10	.25	2.50	5.0
1.5	20	2.25	45.00	30.0
2.5	60	6.25	375.00	150.0
3.5	70	12.25	857.50	245.0
4.5	40	20.25	810.00	180.0
			2090.00	610.0

$$\bar{x} = 610/200 = 3.05$$

$$s = \sqrt{\frac{2090 - 200(3.05)^2}{199}} = \sqrt{1.15326632} = 1.073902385$$

2. The mean is the sum of all observations times frequencies divided by the total number of observations, or $93/15 = 6.2$.

3. The median is the eighth smallest observation, or 6.0.

4. The value having the largest frequency is 7.0 with a frequency of 6. Therefore, the mode is 7.0.

Review Quiz 5-4

1. A frequency distribution in which data are tallied into cells is called a *grouped* frequency distribution.

2. Use the formula

$$\text{cell width} = \frac{\text{upper limit} - \text{lower limit}}{\text{number of cells}}$$

$$= \frac{8.0 - 3.0}{4} = 1.25$$

3. a. Since there are 24 data points, the guidelines tell us that we should use 5 cells.

b.

Cell	Frequency
1	3
2	4
3	8
4	4
5	5

Review Quiz 5-5

1.

Midpoint	Midpoint × Frequency
2.10	4.20
2.30	11.50
2.50	20.00
2.70	10.80
2.90	14.50
	61.00

The mean is $61.00/24 = 2.54$.

2. Since there are 24 observations, the twelfth or thirteenth value is the middle value. The cumulative frequency of cell 2 is 7; the cumulative frequency of cell 3 is 15. Thus, cell 3 contains the median. The median would be estimated by the midpoint of the cell, or 2.50.

3. Cell 3 has the highest frequency. The mode is estimated as the midpoint of this cell, or 2.50.

4. From problem 1, the mean is 2.54.

x	f	x^2	fx^2
2.1	2	4.41	8.82
2.3	5	5.29	26.45
2.5	8	6.25	50.00
2.7	4	7.29	29.16
2.9	5	8.41	42.05
			156.48

$$s = \sqrt{156.48 - 24(2.54)^2/23}$$

$$= .2671589$$

Review Quiz 5-6

1. *True.*

2. *True*. You generally expect a histogram of process output to follow a bell-shaped pattern that gradually tails off on both ends.
3. *False*. The product may have been sorted before the histogram was constructed. Or, due to the fact that data only represent a sample, some nonconforming product might be produced at a later time.
4. *True*.

End of Chapter 5 Quiz

1. *False*. A frequency distribution is a *tabular* summary of data. A histogram is a graphical summary.
2. *True*.
3. *True*. Frequency distributions and histograms are useful tools for helping determine if production can meet specifications by comparing the specifications against the distribution of process output.
4. *False*. Grouped frequency distributions are most useful for large amounts of data.
5. *True*.
6. The cell width is calculated by the following expression:

$$\text{cell width} = \frac{\text{upper limit} - \text{lower limit}}{\text{number of cells}}$$

$$= \frac{55 - 25}{5} = 30/5 = 6$$

7. The number within specifications is the sum of the frequencies between 145 and 160, or $10 + 20 + 40 = 70$. The number outside specifications is the number greater than 160, or $30 + 20 = 50$. Therefore, the percentage out of specification is

$$\frac{50}{70 + 50} = .4167 \text{ or } 41.67\%$$

CHAPTER 6

Review Quiz 6-1

1. *False*. Process capability is the natural variation of the output.
2. *True*.

3. *False*. Since the natural variation falls well within the specifications, you would expect no nonconforming output.
4. *True*.
5. (d) .00033. Since samples of size 5 were used, $d_2 = 2.326$. The standard deviation is estimated by $.00077/2.326 = .00033$.

Review Quiz 6-2

1. *True*.
2. *True*. The *capability* is good if the process can be centered.
3. *False*. They can be if the process is centered and *Cp* is also greater than one.
4. *False*. *Cpk* is the smaller of *Cpl* and *Cpu*.

End of Chapter 6 Quiz

1. *False*. It is measured by comparing the design specifications to the natural variation of the process.
2. *False*. This is the target we desire, provided that the process is also centered.
3. *True*.
4. $Cp = \dfrac{10.50 - 10.38}{6(.04)} = \dfrac{.12}{.24} = 0.5$

   The process is *not* capable of meeting specifications.
5. $Cpu = \dfrac{10.50 - 10.47}{3(.04)} = \dfrac{.03}{.12} = .25$

   This means that the ability to meet the upper specification is very poor.
6. $Cpl = \dfrac{10.47 - 10.38}{3(.04)} = \dfrac{.09}{.12} = .75$

   While better than the upper capability index, the lower specification still cannot be met most of the time.
7. $Cpk = .25$, the smaller of *Cpu* and *Cpl*.

CHAPTER 7

Review Quiz 7-1

1. *False.* SPC definitely needs top management support.
2. *True.*
3. *False.* All assignable causes should be removed before beginning a process capability study.

Review Quiz 7-2

1.
dent	25
surface scratch	13
paint smear	9
trim misalign.	3

2.
dent	25	50%
surface scratch	13	26%
paint smear	9	18%
trim misalign.	3	6%
		100%

3.
			cumulative
dent	25	50%	50%
surface scratch	13	26%	76%
paint smear	9	18%	94%
trim misalign.	3	6%	100%

4.

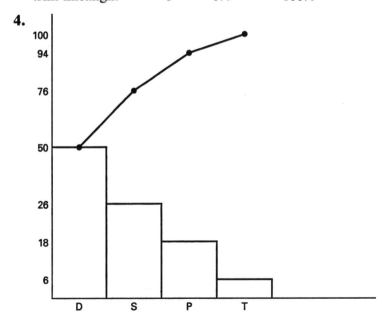

Review Quiz 7-3

1. *False*. It is only a guide to possible reasons. Further study is necessary to verify the cause and effect relationship suggested.
2. (c)
3. *True*.

Review Quiz 7-4

1. *False*. Time is not always one of the variables.
2. (b)
3. (c)

End of Chapter 7 Quiz

1. *False*. Tackle one problem at a time!
2. *True*.
3. *False*. A poor measurement system will not give useful results.
4. (b)
5. (a)
6. *True*.
7. *False*. It does not depend on data, only ideas.

APPENDIX A

Math Skills Analysis

PART 1: BASIC ARITHMETIC

1. $\begin{array}{r} 18 \\ 49 \\ +12 \\ \hline 79 \end{array}$

2. $\begin{array}{r} 33.622 \\ 47.721 \\ 21.276 \\ +68.173 \\ \hline 170.792 \end{array}$

3.   17.837
         −12.527
          5.310

4.   63.721
        − .312
          63.409

5.   23
         ×74
         1702

6.   4,963
          ×1,579
         7,836,577

7. 20)2040 = 102

8. 251)157,628 = 628

PART 2: FRACTIONS

1. 4/8 = 1/2
2. 20/32 = 5/8
3. 15/125 = 3/25
4. 4/5 = 16/20
5. 16/100 = 4/25
6. 1/9 + 3/9 = 4/9
7. 3/5 + 3/4 = 27/20 or 1 7/20
8. 5/7 − 2/7 = 3/7
9. 7/8 − 7/9 = 7/72
10. 2/6 × 3/5 = 1/5
11. 6 × 5/8 = 15/4 or 3 3/4
12. (3/8)/(1/2) = 3/4
13. (10/16)/8 = 5/64

PART 3: DECIMALS

1. 0.9

2. 5.04

3. 3.00077

4. 0.6

5. 7.0016

6. 0.388

7. 3.77

8. .0751

9.   4.072
        +16
        $\overline{}$
        20.072

10.   9.07
        −2.79
        $\overline{}$
        6.28

11.   111.03
        − 6.4
        $\overline{}$
        104.63

12.   .396
        × 16
        $\overline{}$
        6.336

13.   7.69
        × 3.5
        $\overline{}$
        26.915

14. $19\overline{)8.7628}$ = 0.4612

15. $.35\overline{)7763}$ = 22.180

16. 4

17. 576

18. 26.6256

19. 3

20. 0.4

21. 1.5

Review Quiz A-2

1. 5/7
2. The correct answer is (a). The top portion of a fraction is called the numerator.
3. The correct answer is (b). The lower portion of a fraction is called the denominator.
4. 2/3. The fraction is already fully reduced.
5. Divide the numerator and denominator by 3, obtaining 3/4.
6. Divide the numerator and denominator by 7, obtaining 2/5.

Review Quiz A-3

1. 0.2
2. 0.667
3. 0.57

Review Quiz A-4

1. $\begin{array}{r} 4.32 \\ +1.56 \\ \hline 5.88 \end{array}$

2. $\begin{array}{r} 5.60 \\ -3.25 \\ \hline 2.35 \end{array}$

3. $3 \times 3 = 9$
4. $\sqrt{16} = 4$

C

Control Chart Factors

Sample size	A_2	D_3	D_4
2	1.880	0	3.267
3	1.023	0	2.574
4	.729	0	2.282
5	.577	0	2.114
6	.483	0	2.004
7	.419	.076	1.924
8	.373	.136	1.864
9	.337	.184	1.816
10	.308	.223	1.777
11	.285	.256	1.744
12	.266	.283	1.717
13	.249	.307	1.693
14	.235	.328	1.672
15	.223	.347	1.653

Bibliography

AMSDEN, ROBERT T., HOWARD E. BUTLER, AND DAVIDA M. AMSDEN. *SPC Simplified: Practical Steps to Quality*. White Plains, NY: UNIPUB/Kraus International Publications, 1986.

BERGER, ROGER W., AND THOMAS HART. *Statistical Process Control, A Guide for Implementation*. New York: Marcel Dekker, 1986.

BURR, IRVING W. *Elementary Statistical Quality Control*. New York: Marcel Dekker, 1979.

CROSBY, PHILIP B. *Quality is Free*. New York: McGraw-Hill, 1979.

DataMyte Handbook, 3rd Edition. DataMyte Corporation, 14960 Industrial Road, Minnetonka, MN 55345, 1987.

DiPRIMIO, ANTHONY. *Quality Assurance in Service Organizations*. Radnor, PA: Chilton Book Co., 1987.

EVANS, JAMES R., AND WILLIAM M. LINDSAY. *The Management and Control of Quality*. St. Paul: West Publishing Co., 1989.

General Motors Statistical Process Control Manual. American Society for Quality Control, 310 West Wisconsin Ave., Milwaukee, WI 53203, 1986.

GITLOW, HOWARD S., AND SHELLY J. GITLOW. *The Deming Guide to Quality and Competitive Position*. Englewood Cliffs, NJ: Prentice-Hall, 1987.

Glossary and Tables for Statistical Quality Control. American Society for Quality Control, Milwaukee, WI, 1983.

GRIFFITH, GARY. *Quality Technician's Handbook.* New York: John Wiley & Sons, 1986.

GROOCOCK, JOHN M. *The Chain of Quality.* New York: John Wiley & Sons, 1986.

HARRINGTON, H. JAMES. *The Improvement Process: How America's Leading Companies Improve Quality.* New York: McGraw-Hill, 1987.

HRADESKY, JOHN L. *Productivity and Quality Improvement.* New York: McGraw-Hill, 1988.

ISHIKAWA, KAORU. *Guide to Quality Control.* Tokyo: Asian Productivity Organization, 1982.

KRISHNAMOORTHI, K. S. *Quality Control for Operators and Foremen.* Milwaukee, WI: ASQC Quality Press, 1989.

LESTER, RONALD H., NORBERT L. ENRICK, AND HARRY E. MOTTLEY, JR. *Quality Control for Profit, Second Edition.* New York: Marcel Dekker, 1985.

MESSINA, WILLIAM S. *Statistical Quality Control for Manufacturing Managers.* New York: John Wiley & Sons, 1987.

MORSE, WAYNE J., HAROLD P. ROTH, AND KAY M. POSTON. *Measuring, Planning, and Controlling Quality Costs.* Montvale, NJ: National Association of Accountants, 1987.

OAKLAND, JOHN S. *Statistical Process Control, A Practical Guide.* London: William Heinemann Ltd., 1986.

OTT, ELLIS R. *Process Quality Control.* New York: McGraw-Hill, 1975.

SCHROCK, EDWARD M., AND HENRY L. LEFEVRE. *The Good and the Bad News About Quality.* New York: Marcel Dekker, 1988.

SEPHRI, M. *Quest for Quality.* Norcross, GA: Industrial Engineering and Management Press, 1987.

SPECHLER, JAY W. *When America Does It Right: Case Studies in Service Quality.* Norcross, GA: Industrial Engineering and Management Press, 1988.

Statistical Quality Control Handbook. AT&T Technologies, Commercial Sales Clerk, Select Code 700-444, P. O. Box 19901, Indianapolis, IN 46219.

STEBBING, LIONEL. *Quality Assurance: The Route to Efficiency and Competitiveness.* West Sussex, England: Ellis Horwood Ltd., 1986.

WALSH, LOREN, RALPH WURSTER, AND RAYMOND J. KIMBER, EDITORS. *Quality Management Handbook.* New York: Marcel Dekker, 1986.

Index

A

Accuracy, measurement of, 8–9
American Society of Quality Control
     (ASQC), 2, 10, 76
Applications, histograms, 127–30
Arithmetic skills analysis, basic, 157–58
Assignable (special) causes of variation,
     5, 67
Attributes data:
  control charts for, 71–75
  definition, 7
  and frequency distributions, 103
  values assumed by, 71
Average fraction nonconforming, compu-
     tation of, 73
Average range, 30–31, 88
Average value. *See* Mean

B

Basic arithmetic, skills analysis, 157–58
Bell Telephone System, 2
Bi-modal histograms, 107
Blowholes, 103

C

Cause and effect diagrams, 150–52
  constructing, 151
  example of, 152
  and problem solving, 151–52
c-charts, 89–93
  control limits for, 89–90
  and sample sizes, 90
  standard deviation, 89

187